SpringerBriefs in Computer Science

SpringerBriefs present concise summaries of cutting-edge research and practical applications across a wide spectrum of fields. Featuring compact volumes of 50 to 125 pages, the series covers a range of content from professional to academic.

Typical topics might include:

- A timely report of state-of-the art analytical techniques
- A bridge between new research results, as published in journal articles, and a contextual literature review
- A snapshot of a hot or emerging topic
- An in-depth case study or clinical example
- A presentation of core concepts that students must understand in order to make independent contributions

Briefs allow authors to present their ideas and readers to absorb them with minimal time investment. Briefs will be published as part of Springer's eBook collection, with millions of users worldwide. In addition, Briefs will be available for individual print and electronic purchase. Briefs are characterized by fast, global electronic dissemination, standard publishing contracts, easy-to-use manuscript preparation and formatting guidelines, and expedited production schedules. We aim for publication 8–12 weeks after acceptance. Both solicited and unsolicited manuscripts are considered for publication in this series.

**Indexing: This series is indexed in Scopus, Ei-Compendex, and zbMATH **

More information about this series at https://link.springer.com/bookseries/10028

Alessandro Betti · Marco Gori · Stefano Melacci

Deep Learning to See

Towards New Foundations of Computer Vision

 Springer

Alessandro Betti
Université Côte d'Azur, Inria, CNRS,
Laboratoire I3S, Maasai team
Nice, France

Marco Gori ⓘD
Department of Information Engineering
and Mathematics (DIISM)
University of Siena
Siena, Italy

Stefano Melacci
Department of Information Engineering
and Mathematics (DIISM)
University of Siena
Siena, Italy

ISSN 2191-5768   ISSN 2191-5776   (electronic)
SpringerBriefs in Computer Science
ISBN 978-3-030-90986-4   ISBN 978-3-030-90987-1   (eBook)
https://doi.org/10.1007/978-3-030-90987-1

This Springer imprint is published by the registered company Springer Nature Switzerland AG
The registered company address is: Gewerbestrasse 11, 6330 Cham, Switzerland

To All People Who Love To Ask Questions

Preface

DEEP learning has revolutionized computer vision and visual perception. Among others, the great representational power of convolutional neural networks and the elegance and efficiency of backpropagation have played a crucial role. By and large, there is a strong scientific recognition of their popularity, which is very well deserved. However, as yet, most significant results are still based on the truly artificial supervised learning communication protocol, which sets in fact a battlefield for computers, but it is far from being natural. In this book, we argue that, when relying on supervised learning, we have been working on a problem that is—from a computational point of view—remarkably different and like more difficult with respect to the one offered by nature, where motion is in fact in charge for generating visual information. Could not be the case that motion is fact nearly all what we need for learning to see? Otherwise, how could eagles acquire such a spectacular visual skills? What else could they grasp from a video to extract precious information for learning? For sure, just like other animals, they do not undergo a massive supervision, but only a reinforcement signal due to their natural interactions with the environment. Current deep learning approaches based on supervised images mostly neglect the crucial role of temporal coherence. It looks like nature did a nice job by using time to sew all the video frames. When computer scientists began to cultivate the idea of interpreting natural video, in order to simplify the problem they remove time, the connecting wire between frames. As a consequence, video turned into huge collections of images, where temporal coherence was lost, which means that we are neglecting a fundamental clue to interpret visual information, and that we have ended up into problems where the extraction of the visual concepts can only be based on spatial regularities.

Based on the underlying representational capabilities of deep architectures and learning algorithms that are still related to backpropagation, in this book we propose that the massive image supervision can in fact be replaced with the natural communication protocol arising from living in a visual environment, just like animals do. This leads to formulate learning regardless of the accumulation of labeled visual databases, but simply by allowing visual agents to live in their own visual environments. We claim that feature learning arises mostly from motion invariance principles that turns out to be fundamental for detecting the object identity as well as supporting object affordance.

This book introduces two fundamental principles of visual perception. The *first principle* involves consistency issues, namely the preservation of material point identity during motion. Depending on the pose, some of those points are projected onto the retina. Basically, the material points of an object are subject to *motion invariance of the corresponding pixels on the retina*. A moving object clearly does not change its identity, and therefore, imposing an invariance leads to a natural formulation of object recognition. Interestingly, more than the recognition of an object category, this leads to the discovering of its identity.

Motion information does confer not only object identity, but also its affordance, which corresponds with its function in real life. Affordance makes sense for a species of animal, where specific actions take place. A chair, for example, has the affordance of seating a human being, but it can have other potential uses. The *second principle of visual perception* is about its affordance as transmitted by coupled objects— typically humans. The principle states that the affordance is invariant under the coupled object movement. Hence, a chair gains the seating affordance independently of the movement of the person who is sitting (coupled object).

The theory of deep learning to see that is herein proposed is independent of the body of the visual agent since it is only based on information-based principles. In particular, we introduce a vision field theory for expressing those motion invariance principles. The theory enlightens the indissoluble pair of visual features and their conjugated velocities, thus extending the classic brightness invariance principle for the optical flow estimation. The emergence of visual features in the natural framework of visual environments is given a systematic foundation by establishing information-based laws that naturally enable deep learning processes.

The ideas herein presented have been stimulated by a number of questions that we regard of fundamental importance for the construction of a theory of vision. How can animals conquer visual skills without requiring the "intensive supervision" we impose to machines? What is the role of time? More specifically, what is the interplay between the time of the agent and the time of the environment? Can animals see in a world of shuffled frames like computers do? How can we perform semantic pixel labeling by receiving only a few supervisions? Why has the visual cortex evolved toward a hierarchical organization and why did it split into two functionally separated mainstreams? Why top-level visual skills are achieved in nature by animals with foveated eyes thanks to focus of attention? What drives eye movements? Why does it take 8–12 months for newborns to achieve adult visual acuity? How can we develop "linguistic focusing mechanisms" that can drive the process of object recognition?

This book is a humble attempt at addressing these questions, and it is far away from providing a definite answer. However, the proposed theory gives foundations and insights to stimulate future investigations and specific applications to computer vision. Moreover, the field theory herein proposed might also open the doors to disclose interesting problems in visual perception and capture experimental evidence in neuroscience.

Siena, Italy
August 2021

Alessandro Betti
Marco Gori
Stefano Melacci

Acknowledgements

It is hard not to forget people who have contributed in different ways to shape the ideas that are proposed in this book.

First, the research team in Siena Artificial intelligence Lab (SAILab) has played the most important support for maturing the ideas elaborated in this book. Matteo Tiezzi, Dario Zanca, Enrico Meloni, Lapo Faggi, and Simone Marullo, who are some of the members of the SAILab research team in computer vision, have contributed with their experimental results to shape our ideas and make the theory simpler and more general. There are in fact significant traces of early studies in SAILab from the collaboration with Marco Lippi and Marco Maggini, who contributed to develop the first approaches to learning to see by using motion invariance. In particular, discussions with Marco Maggini, who early discovered a number of slippery issues in the incorporation of motion invariance and clearly identified the major difficulties in carrying out learning in the temporal domain, have been extremely inspiring.

The road that has led to this book certainly crosses some inspiring meeting with Marcello Pelillo and, later on, with Fabio Roli. Marcello ignited my latent passion for unifying and that of looking for invariants in vision. Together with Fabio, years ago, we cultivated the dream of a truly new way of facing computer vision challenges within the "en plein air" framework, which reminds us of painting outdoor— machines which learn directly in their own environment. This is in fact addressed at the end of the book. Most of the comments definitely come from early discussions which began with the Workshop GIRPR 2014.

Oswald Lanz has been for us the main reference and source of inspirations for studies in optical flow, which has subsequently given rise to the two principles of perceptual vision. In particular, the conception of the idea of specific velocities associated with visual features has been originated by his clear presentation of the state of the art in optical flow and in tracking.

The studies by Tomaso Poggio on developing visual features under invariance stimulated very fruitful discussion in connection with the Workshop on "Biologically Plausible Learning" at LOD 2020 and have fueled significantly the development of the theory presented in this book.

During the Ph.D. studies of Alessandro Betti, we benefited from a number of constructive criticisms from Stefano Soatto and Michael Bronstein. Among others, they stimulated the importance of setting up an experimental framework adequate to assess the performance. In the same direction, a discussion with Bastian Leibe on the current state of the art in computer vision has contributed to shape and reinforce the idea of "en plein air" discussed in the Epilogue of the book, thus promoting the fundamental principle of replacing the accumulation of visual databases with virtual visual environments. The studies by Ulisse Stefanelli on the reformulation of the principle of least action in physics were a fundamental source of inspiration for the development of the online formulation of learning reported in this book. The same idea used in mechanics gives rise to online gradient-based learning which is nicely related to classic stochastic gradient descent.

We have been inspired by a number of studies in neuroscience, particularly on the mechanisms behind eye movements. Leonardo Chelazzi's visit to SAILab was very influential concerning the subsequent development of computational models of focus of attention. We strongly benefited from discussions on the different kinds of eye movements and, particularly, on the supposed inhibition of video transmission during saccadic movements. The collaboration with Alessandra Rufa offered the primary support on the formulation of theory of gravitational attraction of attention, which is also at the basis of the local spatiotemporal model reported in the book. Giuseppe Boccignone provided a rich bibliographic support and stimulated many discussions mostly on the joint role of action and perception and on the vision blurring process in newborns and in chicks.

This work has been partially supported by the French government, through the 3IA Côte d'Azur, Investment in the Future, project managed by the National Research Agency (ANR) with the reference number ANR-19-P3IA-0002 and by the HumaneAI European research grant Grant agreement ID: 952026 (University of Pisa and University of Siena).

Contents

Chapter 1
Motion Is the Protagonist of Vision

There's not a morning I begin without a thousand questions running through my mind ... The reason why a bird was given wings If not to fly, and praise the sky ...

From Yentl, "Where is it Written?" - I. B. Singer, The Yeshiva Boy

Keywords Motion · Learning protocol · Video streams · Learning in the wild

1.1 Introduction

A good way to approach to fundamental problems is to pose the right questions. Posing the right questions does require a big picture in mind and the need to identify reasonable intermediate steps. When, years ago, we began together the contamination with computer vision, a lot of questions run through our mind that we think make this field really fascinating! First of all, we were wondering whether can we regard computer vision as an application of machine learning. We early came up with the answer that while learning does play a primary role in vision, the current methodologies do not really address its essence yet.[1] The impressive quality of visual skills ordinarily exhibited by humans and animals led us to pose a number of specific questions: How can animals conquer visual skills without requiring the "intensive supervision" we impose on machines? What is the role of time? More specifically, what is the interplay between the time of the agent and the time of the environment? Can animals see in a world of shuffled frames like computers do? How can we perform semantic pixel labeling without receiving any specific single supervision on that task? Why has the visual cortex evolved toward a hierarchical organization and why did it split into two functionally separated mainstreams? Why top-level visual skills are achieved in nature by animals with foveated eyes thanks to focus of attention? What drives eye movements? Why does it take 8–12 months for newborns to

---

[1] The challenge of posing the right questions.

A. Betti et al., *Deep Learning to See*, SpringerBriefs in Computer Science, https://doi.org/10.1007/978-3-030-90987-1_1

achieve adult visual acuity? How can we develop "linguistic focusing mechanisms" that can drive the process of object recognition? In the last few years, these questions have driven our curiosity in the science of vision. In this book, we address those questions while disclosing information-based laws on the emergence of visual features that drive the computational processes of visual perception in case of natural communication protocols.[2] Could not be the case that in computer vision we have been working on a problem that is significantly more difficult than the one offered by nature. This is mostly what we ask in this chapter, where we stimulate the unifying principle of extracting visual information from motion only. This chapter is organized as follows. In the next section, we begin giving a big picture of what this book is about, while in Sect. 1.3 we begin addressing the motivations by stressing the artificial nature of the supervised learning protocol and the involved complexity issues. In Sect. 1.4, we emphasize the importance of cutting the umbilical link with pattern recognition, while in Sect. 1.5 we stress the importance of working with video instead of collections of labeled images. Finally, in Sect. 1.6 we formulate *ten questions* for a theory of vision that drives most of the analyses reported in this book.

1.2   The Big Picture

The remarkable progress in computer vision on object recognition in the last few years achieved by deep convolutional neural networks [62] is strongly connected with the availability of huge labeled data paired with strong and suitable computational resources. Clearly, the corresponding supervised communication protocol between machines and the visual environments is far from being natural. This protocol sets in fact a battlefield for computers. We should not be too surprised that very good results are already obtained, just like the layman is not surprised that computers are quick to do multiplication! Still, one question is in order: Is not the case that while facing nowadays object recognition problems we have been working on a problem that is significantly more difficult than the one offered by nature? Current deep learning approaches based on supervised images mostly neglect the crucial role of temporal coherence. It looks like nature did a nice job by using time to sew all the video frames. When computer scientists began to cultivate the idea of interpreting natural video, in order to simplify the problem they remove time, the connecting wire between frames. As a consequence, video turned into huge collections of images, where temporal coherence was lost. At a first glance, this is reasonable, especially if we consider that, traditionally, video was heavy data sources. However, a closer look reveals that we are in fact mostly neglecting a fundamental clue to interpret visual information, and that we have ended up into problems where the extraction of the visual concepts can only be based on spatial regularities.

---

[2] Could not be the case that in computer vision we have been working on a problem that is significantly more difficult than the one offered by nature?

As we decide to frame visual learning processes in their own natural video environment, we early realize that perceptual visual skills cannot emerge from massive supervision on different object categories. Moreover, linguistic skills arise in children when vision has already achieved a remarkable degree of development and, most importantly, there are animals with excellent visual skills. How can an eagle distinguish a prey at distances of kilometers? The underlying hypothesis that gives rise to the natural laws of visual perception proposed in this book is that visual information comes from motion. Foveated animals move their eyes, which means that even still images are perceived as patterns that change over time. This book arises from the belief that we can understand the underlying information-based laws of visual perception regardless of the body of the agent. No matter whether we are studying chicks, newborns, or computers, the information that arises from the surroundings through the light that enters their "eyes" is the same. As they open their eyes, they need to tackle problems that involve functional issues that are only related to the nature of the source. Since information is interwound with motion, we propose to explore the consequences of stressing the assumption that the focus on motion is in fact *nearly all that we need*. When trusting this viewpoint, one early realizes that, since we involve the optical flow, the underlying computational model must refer to single pixels at a certain time. In nature, best visual perception is in fact achieved by animals which focus attention, a feature that turns out to play a crucial role in terms of computational complexity. We show that this pixel-based assumption leads to state two invariant principles that drive the construction of computational models of visual perception, whenever the agent is supposed to establish natural communication protocols.[3]

A major claim here is that motion is nearly all you need for extracting information from a visual source. We introduce *two principles of visual perception* which express the invariance with respect to the object and to the coupled object motion, respectively. Motion is what offers us an object in all its poses. Classic translation, scale, and rotation invariances can clearly be gained by appropriate movements of a given object. However, the experimentation of visual interaction due to motion goes well beyond the need of these invariances and it includes the object deformation, as well as its obstruction. Only small portions can be enough for the detection, even in the presence of environmental noise. The *first principle of visual perception* involves consistency issues, namely the preservation of material points identity during motion. Depending on the pose, some of those points are projected onto the retina. Basically, the material points of an object are subject to *motion invariance of the corresponding pixels on the retina*. A moving object clearly does not change its identity, and therefore, imposing an invariance leads to a natural formulation of object recognition. Interestingly, more than the recognition of an object category, this leads to the discovering of its identity. Motion information does not only confer object identity, but also its affordance, which corresponds with its function in real life. Affordance makes sense for a species of animal, where specific actions take place. A chair, for example, has the affordance of seating a human being, but it can

---

[3] Motion is nearly all what you need: the two principles of motion invariance.

have other potential uses. The *second principle of visual perception* is about its affordance as transmitted by coupled objects—typically humans. The principle states that the affordance is invariant under the coupled object movement. Hence, a chair gains the seating affordance independently of the movement of the person who is sitting (coupled object).[4]

These two principles drive an information-based approach to visual perception. It is based on a field theory which predicts the visual features, including the object codes, and their associated velocities in all pixels of the retina for any frame. Motion invariance leads to formulate constraint-based computational model where both the velocities and the features are unknown variables, which are determined under the regularization principle which reminds us of predicting coding, where we enforce the reproduction of the visual information from motion invariant features.

The view of visual perception proposed in this book might have an impact on computer vision, where one could open the challenge of abandoning the protocol of massive supervised learning based on huge labeled databases toward the more natural and simple framework in which machines, just like animals, are expected to learn to see in their own visual environment.

1.3 Supervised Learning Is an Artificial Learning Protocol

The professional construction of huge supervised visual databases has significantly contributed to the spectacular performance of deep learning. However, the extreme exploitation of the truly artificial communication protocol of supervised learning has been experimenting its drawbacks, including the vulnerability to adversarial attacks, which might also be tightly connected to the negligible role typically played by the temporal structure of video signals.[5]

At the dawn of pattern recognition, when scientists began to cultivate the idea of interpreting video signals, in order to simplify the problem of dealing with a huge amount of information they removed time, the connecting thread between frames, and began playing with the pattern regularities that emerge at the spatial level. As a consequence, many tasks of computer vision were turned into problems formulated on collections of images, where the crucial role of temporal coherence was neglected. Interestingly, when considering the general problem of object recognition and scene interpretation, the joint role of the computational resources and the access to huge visual databases of supervised data has contributed to erect nowadays "reign of computer vision." At a first glance, this is reasonable, especially if you consider that videos were traditionally heavy data sources to be played with. However, a closer look reveals that we are in fact neglecting a fundamental clue to interpret visual information, and that we have ended up into problems where the extraction of the visual concepts is mostly based on spatial regularities. On the other hand, reigns

---

[4] Motion invariance principles and information-based laws of feature development.

[5] Are not we missing something? It looks like nature did a great job by using time to "sew all the video frames," whereas it goes unnoticed to our eyes!

have typically consolidated rules from which it is hard to escape. This is common in novels and real life. "The Three Princes of Serendip" is the English version of "Peregrinaggio di tre giovani figliuoli del re di Serendippo," published by Michele Tramezzino in Venice on 1557. These princes journeyed widely, and as they traveled they continually made discoveries, by accident and sagacity, of things they were not seeking. A couple of centuries later, in a letter of January 28, 1754, to a British envoy in Florence, the English politician and writer Horace Walpole coined a new term: serendipity, which is succinctly characterized as the art of finding something when searching for something else. Could not similar travels open new scenario in computer vision? Could not the visit to well-established scientific domains, where time is dominating the scene, open new doors to an in-depth understanding of vision? We need to stitch the frames to recompose the video using time as a thread, the same thread we had extracted to work on the images at the birth of the discipline.[6]

This book is a travel toward the frontiers of the science of vision with special emphasis on object perception. We drive the discussion by a number of questions— see Sect. 1.6—that mostly arise as one tries to interpret and disclose natural vision processes in a truly computational framework. Regardless of the success in addressing those questions, this book comes from the awareness that posing right questions on the discipline might themselves stimulate its progress. Abandoning supervised images and letting the agent live in its own visual environment by exploiting the variations induced by motion leads to discover general information-based principles that can shed light also on the neurobiological structure of visual processes. Hopefully, this can also open the doors to an in-depth rethinking of computer vision.

1.4   Cutting the Umbilical Cord with Pattern Recognition

In the eighties, Satosi Watanabe wrote a seminal book [87] in which he early pointed out the different facets of pattern recognition that can be regarded as perception, categorization, induction, pattern recognition, as well as statistical decision making. Most of the modern work on computer vision for object perception fits with Watanabe's view of pattern recognition as statistical decision making and pattern recognition as categorization. Based on optimization schemes with billions of variables and universal approximation capabilities, the spectacular results of deep learning have elevated this view of pattern recognition to a position where it is hardly debatable. While the emphasis on a general theory of vision was already the main objective at the dawn of the discipline [71], its evolution has been mostly marked to significant experimental achievements. Most successful approaches seem to be the natural outcome of a very well-established tradition in pattern recognition methods working on images, which have given rise to nowadays emphasis on collecting big labeled image databases (e.g., [26]).

---

[6] Beyond a peaceful interlude: learning theories based on video more than on images!

In spite of these successful results, this could be the time of an in-depth rethinking of what we have been doing, especially by considering the remarkable traditions of the overall field of vision. In his seminal book, Dyson [28] discusses two distinct styles of scientific thinking: unifying and diversifying. He claimed that most sciences are dominated by one or the other in various periods of their history. Unifiers are people whose driving passion is to find general principles, whereas diversifiers are people with the passion of details. Historical progress of science is often a blend of unification and diversification. Could not it be the right time to exploit the impressive literature in the field of vision to conquer a more unified view of object perception?[7]

In the last few years, a number of studies in psychology and cognitive science have been pushing truly novel approaches to vision. In [56], it is pointed out that a critical problem that continues to bedevil the study of human cognition is related to the remarkable successes gained in experimental psychology, where one is typically involved in simplifying the experimental context with the purpose to discover causal relationships. In so doing, we minimize the complexity of the environment and maximize the experimental control, which is typically done also in computer vision when we face object recognition. However, one might ask whether such a simplification is really adequate and, most importantly, if it is indeed a simplification. Are we sure that treating vision as a collection of unrelated frames leads to simplification of learning visual tasks? In this book, we argue that this is not the case, since cognitive processes vary substantially with changes in context.[8] When promoting the actual environmental interaction, in [56] a novel research approach, called "cognitive ethology," is introduced where one opts to explore first how people behave in a truly naturally situation. Once we have collected experience and evidence in the actual environment, then we can move into the laboratory to test hypotheses. This strongly suggests that also machines should learn in the wild!

Other fundamental lessons come from the school of robotics for whatever involves the control of object-directed actions. In [7], it is pointed out that "the purpose of vision is very different when looking at a static scene with respect to when engaging in real-world behavior." The interplay between extracting visual information and coordinating the motor actions is a crucial issue to face for gaining an in-depth understanding of vision. One early realizes that manipulation of objects is not something that we learn from a picture; it looks like we definitely need to act ourself if we want to gain such a skill. Likewise, the perception of the objects that we manipulate can nicely get a reinforcement from such a mechanical feedback. The mentioned interplay between perception and action finds an intriguing convergence in the natural processes of gaze control and, overall, on the focus of attention [5]. It looks like *animate vision* goes beyond passive information extraction and plays an important role in better posing most vision tasks.[9]

The studies in computer vision might benefit significantly also from the exploration of the links with predictive coding [77] that have had a remarkable impact in

---

[7] From pattern recognition to computer vision.

[8] Cognitive ethology and learning in the wild.

[9] Predictive coding and mental visualization processes.

neuroscience. In that framework, one is willing to study theories of brain in which it constantly generates and updates a "mental model" of the environment. Overall, the model is supposed to generate its own predictions of sensory input and compare them to the actual sensory input. The prediction error is expected to be used to update and revise the mental model. While most of the studies in deep learning have been focused on the direct learning of object categories, there are a few contributions also in the direction of performing a sort of predictive coding by means of auto-encoding architectures [89]. In this book, we go one step beyond by enforcing the extraction of motion independent features. The corresponding invariance is the secrete for identifying the objects as well as the affordance. We shall see that the extracted codes have an inherently symbolic nature which makes it possible to think of images and consequently to carry out the mental visualization processes.

1.5   Dealing with Video Instead of Images

While the emphasis on methods rooted on still images is common on most of nowadays state-of-the-art object recognition approaches, in this book we argue that there are strong arguments to start exploring the more natural visual interaction that animals experiment in their own environment. The idea of shifting to video is very much related to the growing interest of *learning in the wild* that has been explored in the last few years.[10] The learning processes that take place in this kind of environments have a different nature with respect to those that are typically considered in machine learning. While ImageNet [26] is a collection of unrelated images, a video supports information only when motion is involved. In the presence of still images that last for awhile, the corresponding stream of equal frames only conveys the information of a single image—apart from the duration of the interval in which the video has been kept constant. As a consequence, visual environments mostly diffuse information only when motion is involved. As time goes by, the information is only carried out by motion, which modifies one frame to the next one according to the optical flow. Once we deeply capture this fundamental feature of vision, we realize that a different theory of machine learning is needed that must be capable of naturally processing streams instead of a collection of independent images.

The scientific communities of pattern recognition and computer vision have a longstanding tradition of strong intersections. As the notion of pattern began to circulate at the end of the fifties, the discipline of pattern recognition grew up quickly around it. No matter what kind of representation a pattern is given, its name conveys the underlining assumption that, regardless of the way they are obtained, the recognition process involves single entities. It is in fact their accumulation in appropriate collections which gives rise to statistical machine learning.[11] While computer

---

[10] See., e.g., https://sites.google.com/site/wildml2017icml/.

[11] The dominant foundations are still those promoted by the pioneers of machine learning and pattern recognition.

vision can hardly be delimited in terms of both methodologies and applications, the dominant foundations are still those promoted by the pioneers of pattern recognition. Surprisingly enough, there is no much attention on the crucial role of the position (pixel) on which the decision is carried out and, even more, the role of time in the recognition processes does not seem to play a central role. It looks like we are mostly ignoring that we are in front of spatiotemporal information, whose reduction to isolated patterns might not be a natural solution.

In the last decades, the massive production of electronic documents, along with their printed version, has given rise to specialized software tools to extract textual information from optical data. Most optical documents, like tax forms or invoices, are characterized by a certain layout which dramatically simplifies the process of information extraction. Basically, as one recognizes the class of a document, its layout offers a significant prior on what we can expect to find in its different areas. For those documents, the segmentation process can often be given a somewhat formal description, so as most of the problems are reduced to deal with the presence of noise. As a matter of fact, in most real-world problems, the noise does not compromise significantly the segmentation that is very well driven by the expectations provided in each pixel of the documents. These guidelines have been fueling the field of documental analysis and recognition (DAR), whose growth in the last few years has led to impressive results [70]. Unfortunately, in most real-world problems, as we move to natural images and vision, the methodology used in DAR is not really effective in most challenging problems. The reason is that there is no longer a reliable anchor to which one can cling for segmenting the objects of a scene. While we can provide a clear description of chars and lines in optical documents, the same does not hold for the picture of a car which is mostly hidden by a truck during the overtaking. Humans exhibit a spectacular detection ability by simply relying on small glimpses at different scales and rotations. In no way those cognitive processes are reducible to the well-posed segmentation problems of chars and lines in optical documents. As we realize that there is a car, we can in fact provide its segmentation. Likewise, if an oracle gives us the segmented portion of a car, we can easily classify it. Interestingly, we do not really know which of the two processes is given a priority—if any.[12] We are trapped into the *chicken–egg dilemma* on whether classification of objects must take place first of segmentation or vice versa. Among others, this issue has been massively investigated in [19] and early pointed out in [84]. This intriguing dilemma might be connected with the absence of focus of attention, which necessarily leads to holistic mechanisms of information extraction. Unfortunately, while holistic mechanisms are required at a certain level of abstraction, the segmentation is a truly local process that involves also low-level features.

The bottom line is that most problems of computer vision are posed according to the historical evolution of the applications more than according to an in-depth analysis of the underlying computational processes. While this choice has been proven to be successful in many real-world cases, stressing this research guideline might lead, on the long run, to sterile directions. Somewhat outside the mainstream of massive

---

[12] The chicken–egg dilemma.

Table 1.1 Reference to the section where we address the ten questions throughout the book

Question	Q1	Q2	Q3	Q4	Q5	Q6	Q7	Q8	Q9	Q10
Sections	1.2	1.4	3.1	2.1	2.2	3.1	2.2	2.4	5.2	6
	1.3	2.2			4.1	3.4	2.3		5.5	
					4.3		2.4			
							2.5			

exploration of supervised learning, Poggio and Anselmi [76] pointed out the crucial role of incorporating appropriate visual invariance into deep nets to go beyond the simple translation equivariance that is currently characterizing convolutional networks. They propose an elegant mathematical framework on visual invariance and enlighten some intriguing neurobiological connections. Could not it be the case that the development of appropriate invariances is exactly what we need to go one step beyond and also to improve significantly the performance on object recognition tasks?

1.6 Ten Questions for a Theory of Vision

A good way to attack important problems is to pose the right question. John Tukey, who made outstanding contributions on the Fourier series, credited for the invention of the term "bit," during his research activity underlined the importance of posing appropriate questions for an actual scientific development. In his own words:

> Far better an approximate answer to the right question, which is often vague, than the exact answer to the wrong question, which can always be made precise.

Overall, posing appropriate questions can open a debate and solicit answers.[13] However, the right questions cannot be easily posed since, while they need a big pictures in mind, they also need the identification of reasonable intermediate steps. It is often the case that while addressing little problems, inconsistencies arise that suggest the formulation of better questions. Here, we formulate ten questions on the emergence of visual skills in nature that might also contribute to the development of new approaches to computer vision that is based in processing of video instead of huge collections of images. While this book is far from purporting to provide definitive answers to those questions, it contains some insights that might stimulate an in-depth rethinking mostly of object perception. Moreover, it might also suggest different research directions in the control of object-directed action.

Q1 How can animals conquer visual skills without requiring "intensive supervision"?

---

[13] The ten driving questions.

In the last few years, the field of computer vision has been at the center of a revolution. Millions of supervised images have driven a substantial change in the methodology behind object recognition. No animal in nature can compete with machines on the game of supervised learning. Yet, animals don't rely on linguistic supervision. Similarly, humans conquer visual skills without such a boring communication protocol! There must be some important reasons behind this fundamental difference.

Q2 How can animals gradually conquer visual skills in their own environments?

Based on their genetic heritage, animals learn to see while living. Machines share some analogies: First, the transfer of visual features learned in different experimental domains somewhat reminds us the importance of genetic heritage in nature. Second, the artificial learning processes, which are based on the weight updating, also remind us the gradual process of learning that we observe in nature. However, this analogy is only apparent, since nowadays deep learning carries out an optimization process whose iteration steps don't correspond with the environmental time that animals experiment during their life. It looks like, machine learning is neglecting the notion of time in a process which is inherently driven by time!

Q3 Could children really acquire visual skills in such an artificial world, which is the one we are presenting to machines? Do not shuffled visual frames increase the complexity of learning to see?

Visual image databases can be thought of the outcome of shuffling the frames of a video. We should pay attention to this issue, which likely surprises the layman, whereas it doesn't seem to attract significant attention in the scientific community. Couldn't be the case that we have been facing a problem more difficult than the one offered by Nature? ImageNet [26] and related visual databases somehow correspond with the permutation of visual sources.

Q4 How can humans exhibit such an impressive skill of properly labeling single pixels without having received explicit pixel-wise supervisions? It is not the case that such a skill is a sort of "visual primitive" that cannot be ignored for efficiently conquering additional skills on object recognition and scene interpretation?

When the emphasis is on applications involving object recognition, we don't necessarily need to process information at pixel level. However, there are cases, like medical image segmentation, where pixel-based computation is required. Interestingly, humans exhibit great performance at semantic pixel labelling. One might wonder whether such a human skill is in fact a fundamental primitive needed for an in-depth understanding of vision. Moreover, as it will become clear in the rest of the book, the construction of a theory on perceptual vision very much benefits from the local reference to single pixels of both optical flow and visual features. The focus on pixel-based features and velocities somewhat leads to interpret vision in terms of a field theory.

Q5 Why are the visual mainstreams in the brain of primates organized according to a hierarchical architecture with receptive fields? Is there any reason why this solution has been developed in biology?

This seems to be one of the secretes of the success of deep convolutional nets, that are currently learning under the supervised protocol. Interestingly, it looks like the hierarchical structure, along with the concept of receptive field are inherently playing a crucial role in biology and in machines. However, one might wonder whether the typical weight sharing, that is used for capturing the translation equivariance is really needed.

Q6 Why are there two different mainstreams in the primates' brain? What are the reasons for such a different neural evolution?

> The answer to this question is not only of interest in neuroscience. It seems reasonable to invoke functional specialization of the brain with distinct areas dedicated to perception and action, respectively. There should be some fundamental difference for those neurons to be allocated separately. Capturing such a difference would likely help the development of computer vision systems.

Q7 Why do primates and other animals focus attention, whereas others, like the frog, do not? Why are diurnal visual skills best achieved by foveated eyes? Why has not evolution led to develop human retina as a uniform structure composed of cones only—the higher-resolution cells—instead of the variable resolution structure that is mostly based on rods? Apart from anatomical reasons, why has a non-uniform resolution been developed?

> More than sixty years ago, the visual behavior of the frog was posing an interesting puzzle [64] which is mostly still on the table. The frog will starve to death surrounded by food if it is not moving! His choice of food is determined only by size and movement, but he cannot perceive information in still images. Interestingly, the frog doesn't focus attention like humans and most primates, who can very well perceive also information in still images. Again, it could be the case that this different visual skills is rooted in fundamental information-based principles and that, in this case, the act of focus of attention plays a crucial role.

Q8 What are the mechanisms that drive eye movements?

> The mechanisms behind eye movements are likely optimizing the acquisition of visual information. This is an exciting challenge, especially if you consider that the process of eye movement is interwound with learning. While in the early stage of life, the information is mostly in details and moving objects, as time goes by, humans focus on specific meaningful objects. While focus of attention helps object perception, it's quite obvious that also the opposite holds true, so as eye movements are likely driven by an intriguing computational loop.

Q9 Why does it take 8–12 months for newborns to achieve adult visual acuity? Is the development of adult visual acuity a biological issue or does it come from higher-level computational laws of vision?

> The conquering of appropriate strategies for driving eye movements seems to expose a facet of a learning behavior that has been observed in newborns. Couldn't be the case that such a filtering is motivated by the need of protection against information overloading? As newborns open their eyes they experiment a visual stream where information extraction does require to capture very complex spatiotemporal regularities. It looks like such a process is based on a sort of equilibrium so as children react by appropriate visual skills to properly smoothed visual streams. Such an equilibrium can hardly be associated with nowadays dominating gradient-based computations.

Q10 How can we develop "linguistic focusing mechanisms" that can drive the process of object recognition?

The computational processes of vision that take place in nature can hardly be understood until we decouple them from language. The massive adoption of supervised learning seems to go exactly in the opposite direction! Objects are perceived by linguistic supervision, whereas vision in nature takes place mostly by unsupervised learning. Linguistic interactions can definitely support the development of specific human visual competence, but they likely help also at developmental stages in which most important and fundamental visual skills have been already achieved.

The rest of the book is mostly a travel driven by the need to address these questions. The pixel-wise computational processes stimulated in **Q4** along with **Q3** on the need to introduce the natural notion of time suggest splitting the analysis in two parts dedicated to focus of attention mechanisms and the investigation of motion invariance principle. The corresponding conception of information-based laws of perception discussed on Chap. 5, along with the study of focus of attention and motion invariance, leads to address all the above questions.

Chapter 2
Focus of Attention

The frog does not seem to see or, at any rate, is not concerned with the detail of stationary parts of the world around him. He will starve to death surrounded by food if it is not moving. His choice of food is determined only by size and movement.

Lettvin et al. 1959

Keywords Focus of attention · Eye movements · Gravitational model of attention

2.1 Introduction

Why does it happen? This is definitely a curious behavior, a somewhat surprising deficit for us to understand. The frog can successfully catch flying insects, but when it is served with the food in an appropriate bowl starves to death! Hence, she does not see still images, which suggests that their interpretation is a more complex visual task than in case of moving objects. In order to address this puzzle, we start addressing question no. 4 on human capability of performing pixel semantic labeling. This propagates the interest to the exploration of the visual skills of foveated animals and to the mechanisms of focus of attention that are at the basis of the vision field theory herein discussed.[1] The velocity of the point where the agent focuses its attention is in fact the first fundamental field that is involved in the proposed information-based laws of perceptual vision. We begin the study of this field simply because it enables a full motion-based interaction in foveate-based animals that is conjectured to be of fundamental importance also in machines. As we assume that there is an underlying process of eye movements, we recognize the importance of continuously interacting with motion fields, even in the case of still images. As it will be pointed out in this chapter, this is in fact a remarkable difference with respect to the visual process taking place in animals that like frogs are based on significantly less effective visual skills. In the presence of eye movements with focus of attention, everything

---

[1] The velocity of the point of focus of attention is the first field of the theory

© The Author(s), under exclusive license to Springer Nature Switzerland AG 2022
A. Betti et al., *Deep Learning to See*, SpringerBriefs in Computer Science,
https://doi.org/10.1007/978-3-030-90987-1_2

is moving. The external motion, which comes either in case of moving objects or in case of moving agent, is integrated with the internal motion of the agent (i.e., eye and/or head movements). As it will be seen in the rest of the book, this is of crucial importance in order to gain high-level perceptual skills. The chapter is organized as follows. In the next section, we discuss how can humans perform pixel semantic labeling, which directly leads us to consider the capability of focusing attention. In Sect. 2.3, we give some insights of the computational processes associated with focus of attention that are based on the evolution of the animal visual system. In Sect. 2.4, we address the issue of the reasons why we focus attention, and finally, in Sect. 2.5 we study the driving mechanisms beyond eye movements.

2.2   How Can Humans Perform Pixel Semantic Labeling?

Many computer vision tasks still rely on the pattern model that is based on an opportune preprocessing of video information represented by a vector. Surprisingly enough, state-of-the-art approaches to object recognition already offer quite accurate scene descriptions in specific real-world contexts, without necessarily relying on the semantic labeling of each pixel. A global computational scheme emerges that is typically made more and more effective when the environment in which the machine is supposed to work is quite limited, and it is known in advance. In object recognition tasks, the number of the classes that one expects to recognize in the environment can dramatically affect the performance.[2] Interestingly, very high accuracy can be achieved without necessarily being able to perform the object segmentation and, therefore, without needing to perform pixel semantic labeling. However, for an agent to conquer visual capabilities in a broad context, it seems to be very useful to rely on appropriate primitives. We humans can easily describe a scene by locating the objects in specific positions, and we can describe their eventual movement. This requires a deep integration of visual and linguistic skills that are required to come up with compact, yet effective video descriptions. Humans' semantic pixel labeling is driven by the focus of attention that is at the core of all important computational processes of vision. While pixel-based decisions are inherently interwound with a certain degree of ambiguity, they are remarkably effective. The linguistic attributes that we can extract are related to the context of the pixel that is taken into account for label attachment, while the ambiguity is mostly a linguistic more than a visual issue. In a sense, this primitive is fundamental for conquering higher abstraction levels. How can this be done? The focus on single pixels allows us to go beyond object segmentation based on sliding windows. Instead of dealing with object proposals [95], a more primitive task is that of attaching symbols to single pixels in the retina. The bottom line is that human-like linguistic descriptions of visual scenes are gained on top of pixel-based features that, as a by-product, must also allow us to perform semantic labeling. The task of semantic pixel labeling leads to process the retina by focusing

---

[2] Semantic pixel labeling as a fundamental visual primitive.

attention on the given pixel, while considering the information in its neighborhood. This clearly opens the doors to an in-depth rethinking of computational processes of vision. It is not only the frame content, but also where we focus attention in the retina that does matter.

Human ability of exhibiting semantic labeling at pixel level is really challenging. The visual developmental processes conquer this ability nearly without pixel-based supervisions. It seems that such a skill is mostly the outcome of the acquisition of the capability to perform object segmentation. This is obtained by constructing the appropriate memberships of the pixels that define the segmented regions. When thinking of the classic human communication protocols, one early realizes that even though it is rare to provide pixel-based supervision, the information that is linguistically conveyed to describe visual scenes makes implicit reference to the focus of attention. This holds regardless of the scale of the visual entity being described. Hence, the emergence of the capability of performing pixel semantic label seems to be deeply related to the emergence of focus of attention mechanisms. The most striking question, however, is how can humans construct such a spectacular segmentation without a specific pixel-based supervision! Interestingly, we can focus on a pixel and attach meaningful labels, without having been instructed for that task.[3]

The primitive of pixel semantic labeling is likely crucial for the construction of human-like visual skills. There should be a hidden supervisor in nature that, so far, has nearly been neglected. We conjecture that it is the optical flow which plays the central role for object recognition. The decision on its recognition must be invariant under motion, a property that does require a formulation in the temporal direction. The capability of focusing on (x, t) seems to break the circularity of the mentioned chicken–egg dilemma. As it will be seen in the following, the local reference to pixels enables the statement of visual constraints on motion invariance that must be jointly satisfied.

2.3 Insights from Evolution of the Animal Visual System

It is well known that the presence of the fovea in the retina leads to focus attention on details in the scene.[4] Such a specialization of the visual system is widespread among vertebrates, and it is present in some snakes and fishes, but among mammals is restricted to haplorhine primates. In some nocturnal primates, like the owl monkey and in the tarsier, the fovea is morphologically distinct and appears to be degenerate. Owl monkey's visual system is somewhat different from other monkeys and apes.

---

[3] **Q4:** Why semantic pixel labeling?

[4] Mammals and haplorhine primates

As its retina develops, its dearth of cones and its surplus of rods mean that this focal point never forms. Basically, a fovea is most often found in diurnal animals, thus supporting the idea that it is supposed to play an important role for capturing details of the scene [78]. But why have not many mammals developed such a rich vision system based on foveated retinas? Early mammals, which emerged in the shadow of the dinosaurs, were likely forced to nocturnal lives, so as to avoid to become their prey [81].[5] In his seminal monograph, Walls [86] conjectured that there has been a long nocturnal evolution of mammals' eyes, which is the reason of the remarkable differences with respect to those of other vertebrates. The idea became known as the "nocturnal bottleneck" hypothesis [35]. Mammals' eyes tend to resemble those of nocturnal birds and lizards, but this does not currently hold for humans and closely related monkeys and apes. It looks they re-evolved features useful for diurnal living after they abandoned a nocturnal lifestyle upon dinosaur extinction. It is worth mentioning that haplorhine primates are not the only mammals which focus attention in the visual environment. Most mammals have quite a well-developed visual system for dealing with details. For example, it has been shown that dogs possess quite a good visual system that share many features with those of haplorhine primates [6]. A retinal region with a primate fovea-like cone photoreceptor density has been identified but without the excavation of the inner retina. Similar anatomical structure observed in rare human subjects has been named fovea plana. Basically, the results found in [6] challenge the dogma that within the phylogenetic tree of mammals, haplorhine primates with a fovea are the sole lineage in which the retina has a central bouquet of cones.[6] In non-primate mammals, there is a central region of specialization, called *area centralis*, which is often located close to the optic axis and demonstrates a local increase in photoreceptor and retinal ganglion cell density that plays a somehow dual role with respect to the fovea. Like in haplorhine primates, in those non-primate mammals we experiment focus of attention mechanisms that are definitely important from a functional viewpoint.

This discussion suggests that the evolution of animals' visual system has followed many different paths that, however, are related to focus of attention mechanisms, that are typically more effective for diurnal animals. There is, however, an evolution path which is definitely set apart, in which the frog is most classic representer.[7] More than sixty years ago, the visual behavior of the frog was posing an interesting puzzle [64] which is mostly still on the table. In the words of the authors:

> The frog does not seem to see or, at any rate, is not concerned with the detail of stationary parts of the world around him. He will starve to death surrounded by food if it is not moving. His choice of food is determined only by size and movement.

No mammal experiment shows such a surprising behavior! However, the frog is not expected to eat like mammals. When tadpoles hatch and get free, they attach themselves to plants in the water such as grass weeds and cattails. They stay there

---

[5] The long nocturnal evolution of mammals' eyes and the bottleneck hypothesis

[6] Fovea versus area centralis

[7] The frog dilemma

for a few days and eat tiny bits of algae. Then, the tadpoles release themselves from the plants and begin to swim freely, searching out algae, plants, and insects to feed upon. At that time, their visual system is ready. Their food requirements are definitely different from what mammals need and their visual system has evolved accordingly for catching flying insects. Interestingly, unlike mammals, the studies in [64] already pointed out that the frogs' retina is characterized by uniformly distributed receptors with neither fovea nor area centralis. Interestingly, this means that the frog does not focus attention by eye movements.[8]

One can easily argue that any action that animals carry out needs to prioritize the frontal view. On the other hand, this leads to the detriment of the peripheral vision that is also very important. In addition, this could apply for the dorsal system whose neurons are expected to provide information that is useful to support movements and actions. Apparently, the ventral mainstream, with neurons involved in the "what" function, does not seem to benefit from foveated eyes. Nowadays, most successful computer vision models for object recognition, just like frogs, use a uniformly distributed retina and do not focus attention. However, unlike frogs, machines seem to conquer human-like recognition capabilities on still images.[9] Interestingly, unlike frogs, nowadays machines recognize quite well food properly served in a bowl. These capabilities might be due to the current strong supervised communication protocol. Machines benefit from tons of supervised input/output pairs, a process which, as already pointed out, cannot be sustained in nature. On the other hand, as already pointed out, in order to attack the task of understanding what is located in a certain position, it is natural to think of eyes based on fovea or on area centralis. The eye movements with the corresponding trajectory of the focus of attention (FOA) are also clearly interwound with the temporal structure of video sources. In particular, humans experiment eye movements when looking at fixed objects, which means that they continually experiment motion. Hence, also in case of fixed images, conjugate, vergence, saccadic, smooth pursuit, and vestibulo-ocular movements lead to acquire visual information from relative motion. We claim that the production of such a continuous visual stream naturally drives feature extraction, since the corresponding convolutional filters, charged of representing features for object recognition, are expected to provide consistent information during motion. The enforcement of this consistency condition creates a mine of visual data during animal life! Interestingly, the same can happen for machines. Of course, we need to compute the optical flow at pixel level so as to enforce the consistency of all the extracted features. Early studies on this problem (see, e.g., [48]), along with recent related improvements (see, e.g., [4]), suggest to determine the velocity field by enforcing brightness invariance.[10] As the optical flow is gained, it can be used to enforce motion consistency on the visual features. These features can be conveniently combined and used to recognize

---

[8] The frog has got uniformly distributed receptors with neither fovea nor area centralis

[9] Unlike frogs, convolutional deep nets recognize the food served in the bowl, but only thanks to the truly artificial supervised learning protocol!

[10] Could not focus of attention involve motion processes capable of generating a mine of supervised information?

objects. Early studies driven by these ideas are reported in [42], where the authors propose the extraction of visual features as a constraint satisfaction problem, mostly based on information-based principles and early ideas on motion invariance.

2.4   Why Focus of Attention?

In[11] this section, we mostly try to address question no. 7 on the reason why some animals focus attention. As already pointed out, first of all, it looks like focus of attention creates a uniform framework in which visual information comes from motion. There are a number of surprising convergent issues that strongly support the need for focus of attention mechanisms for conquering top-level visual skills that is typical of diurnal animals. Basically, it looks like we are faced with functional issues which mostly obey information-based principles that hold regardless of the body of the agent.[12]

1. *The FOA drives the definition of visual primitives at pixel level*
   The already mentioned visual skill that humans possess to perform pixel semantic labeling clearly indicates their capability of focusing on specific points in the retina with high resolution. Hence, FOA is needed if we want to perform such a task. Another side of the coin does reflect the underlying assumption of understanding perceptual vision on the basis of a field theory, where features and velocities are indissolubly paired.

2. *Variable resolution retina*
   Once we focus on a certain pixel, the computation of the associated visual fields, which might be semantically interpreted, benefits from a variable spatial resolution that is decreasing as we move far away from the point of the focus. This seems to be quite a controversial issue. One can easily argue that given the same overall resolution, while we see better close to the focus, we have lower peripheral visual skills. However, this claim makes sense if we consider only the process which takes place on single frames and we disregard the temporal dimension. Basically, visual processes in nature come with a certain velocity, an issue which makes it possible to fool human eyes in the video production. Hence, if we restrict the frame rate to classic cinema ratio (24 frames/s), then a more dense presence of cones, where the eye is focusing attention, makes it possible to use acquisitions at high resolution in different regions of the retina as a consequence of saccadic movements. It looks like we need a trade-off between the velocity of the scan path in focus of attention and the spatial distribution of the resolution. Clearly, as we increase the degree of different resolutions in the retina we need to increase consequently the velocity of the scan paths. While this discussion offers a convincing description of the appropriate usage of the different resolutions, it does

---

[11] Focus of attention as an information-based process

[12] Eight reasons for FOA

not fully address **Q7** on why has not evolution increased the resolution also at the periphery in foveated animals. This issue will be addressed in Sect. 6.1.

3. *Eye movements and FOA help estimating the probability distribution on the retina*
   At any time, a visual agent clearly needs to possess a good estimation of the probability distribution over the pixels of the retina. This is important whenever we consider visual tasks for which the position does matter. This involves both the *where* and *what* neurons of the dorsal and ventral mainstream [41]. In both cases, it is quite obvious that any functional risk associated with the given task should avoid reporting errors in regions of the retina where there is a uniform color. The probability distribution is of fundamental importance, and it is definitely related to saliency maps that can be gained by FOA.

4. *Eye movements and FOA: There is always motion!*
   The interplay between the FOA and motion invariance properties is the key for understanding human vision and general principles that drive object recognition and scene interpretation. In order to understand the nice circle that is established during the processes of learning in vision, let us start exploring the very nature of eye movements in humans. Basically, they produce visual sequences that are separated by saccadic movement, during which no information is acquired. In the case of micro-saccades, the corresponding micro-movements explore regions with a remarkable amount of details that are somehow characterized by certain features. The same holds true for smooth pursuit. No matter what kind of movement is involved, apart from saccadic movements, the visual information is always paired with motion, which leads to impose invariances on the extracted feature.[13] This indicates a remarkable difference between different animals; the frog does not produce such a movement in the presence of still images! It is in fact the presence of eye movements with FOA which somehow unifies the visual interactions, since we are always in the presence of motion. Clearly, the development of such a motion invariance in this case does exploit much more information and can originate skills that are absent in some animals. Basically, the focus of attention mechanisms originates an impressive amount of information that, in this case, nature offers for free! As it will be pointed out in Chap. 3, this is very important for the information that arises from motion invariance principles.

5. *Saccades, visual segments, and "concept drift"*
   The saccadic movements contribute to perform "temporally segmented computations" over the retina on the different sequences produced by micro-saccadic movements. When discussing motion invariance, we will address the problems connected with *concept drift*, where an object is slightly transformed into another one. Clearly, this could dramatically affect the practical implementation of the motion invariance. For example, the child's teddy bear could be slowly transformed into a nice dog. Preserving motion consistency would lead to confuse bears with dogs! However, among different types of FOA trajectories, the saccadic movements play the fundamental role of resetting the process, which clearly faces directly problems of concept drift.

---

[13] An in-depth coverage of this issue is in Chap. 3.

6. *FOA helps disambiguation at learning time*

A puzzle is offered at learning time when two or more instances of the same object are present in the same frame, maybe with different poses and scales. The FOA in this case helps disambiguating the enforcement of motion invariance. While the enforcement of weight sharing is ideal for directly implementing translation equivariance, such a constraint does not facilitate other more complex invariances that can better be achieved by its removal. As it will be more clear in Chap. 5, the removal of the weight constraint adds one more tensorial dimension with significant additional space requirements. However, we shall see appropriate mechanisms for efficiently facing this. Basically, the constraint of motion invariance under FOA movements naturally disambiguates the processing on different instances of the same object since the computation takes place, at any time, on a single instance.

7. *FOA drives the temporal interpretation of scene understanding*

Animals which focus attention receive information from the environment which is in fact driven and defined by the FOA. In particular, humans acquire information on the object(s) on which they are focusing attention. This is a simple, yet powerful mechanism for receiving supervision. Without such pointing mechanism, information on the current frame cannot refer to a specific position, which makes learning more difficult. The importance of FOA involves also the interpretation of visual scenes. It is in fact the way FOA is driven which sequentially selects the information for the scene interpretation. Depending on the purpose of the agent and on its level of scene understanding the FOA is consequently driven. This process clearly shows the fundamental role on the selection of the points where to focus attention, an issue which is described in the following section. Once again, we are back to the issue of the limited number of frames/sec which characterizes natural video. The scene interpretation has its own dynamics which very well fits the corresponding mechanisms of focus of attention.

8. *FOA helps disambiguating illusions*

Depending on where an agent with foveated eyes focuses attention concepts that, strictly speaking, do not exist can emerge, thus originating an illusion. A noticeable example is Kanizsa's triangle, but it looks like other illusions arise for related reasons. You can easily experiment that as you approach any detail, it is perfectly perceived without any ambiguity. A completion mechanism arises that leads us to perceive the triangle as soon as we move away from the figure and the mechanism is favored by focusing attention on the barycenter. Interestingly, the different views coming from different points where an agent with foveated eyes focuses attention likely help disambiguating illusions, a topic that has been recently studied in classic convolutional networks [3, 55]. Kanizsa's triangle is not a special case of illusion. There is a huge literature on related cases, and even Barber's pole [33] that is discussed in the following is an intriguing example of the role of focus of attention in the different interpretations that can be offered of the same object.

The analysis[14] on foveated-based neural computation helps understanding also the reason why humans cannot see video with a number of frames per second that exceeds the classic sampling threshold. It turns out that this number is clearly connected with the velocity of the scan paths of the focus of attention. Of course, this is a computational issue which goes beyond biology and clearly affects machines as well.

> The above items provide strong evidence on the reasons why foveated eyes turn out to be very effective for scene understanding. Interestingly, we can export the information-based principle of focusing attention to computer retinas by simulating eye movements. There is more: Machines could provide multiple focuses of attention which could increase their visual skills significantly.

2.5   What Drives Eye Movements?

Foveated animals need to move their eyes to properly focus attention. Human eyes make jerky saccadic movements during ordinary visual acquisition. One reason for these movements is that the fovea provides high resolution in portions of about $1, 2°$.[15]

Because of such a small high-resolution portions, the overall sensing of a scene does require intensive movements of the fovea. Hence, the fovea movements do represent a good alternative to eyes with uniformly high-resolution retina. The information-based principles discussed so far lead us to conclude that foveated retinas with saccadic movements are in fact a solution that is computationally sustainable and very effective. Fast reactions to changes in the surrounding visual environment require efficient attention mechanisms to reallocate computational resources to most relevant locations in the visual field.

Visual attention plays a central role in our daily activities. While we are playing, teaching a class, or driving a vehicle, the amount of information our eyes collect is way greater than what we are able to process [1, 58]. To work properly, we need a mechanism that only locates the most relevant objects, thus optimizing the computational resources [85]. Human visual attention performs this task so efficiently that, at conscious level, it goes unnoticed.

Attention mechanisms have been the subject of massive investigation also in machines, especially whenever they are asked to solve tasks related to human perception such as video compression, where loss of quality is not perceivable by viewers [44, 50], or caption generation [24, 66]. Following the seminal works by Treisman

---

[14] Question no. 7 on motivations for focus of attention

[15] A quick tour in the literature

et al. [82, 83] and Koch and Ullman [57], as well as the first computational implementations [51], over the last three decades scientists have presented numerous attempts to model FOA [20]. The notion of *saliency map* has been introduced, which consists of a spatial map that indicates the probability of focusing on each pixel. However, attention is essentially a dynamical process and neglecting the temporal dimension may critically lead to a poor description of the phenomenon [17]. Under the *centralized saliency map hypothesis*, shifts in visual attention can be generated through a winner-take-all mechanism [57], selecting the most relevant location in space at each time step. Still, the temporal dynamics of the human visual selection process is not considered. Some authors have tried to fill this gap with different approaches, for example, by preserving the centrality of the saliency map introducing a handcrafted human bias to choose subsequent fixations [61]. Similarly, [53] tries to formalize the idea that during visual exploration top-level cues continue to increase their importance, to the disadvantage of more perceptive low-level information. In [54], the authors propose a bio-inspired visual attention model based on the pragmatic choice of certain proto-objects and learning the order in which these are attended. All of these approaches still assume the centrality of a saliency map to perform a long stack of global computations over the entire visual field before establishing the next fixation point. This is hardly compatible with what is done by humans where, most likely, attention modulates visual signals before they even reach cortex [22, 73] and restricts computation to a small portion of the available visual information [79, 83]. In [18], the gaze shift is modeled as the realization of a stochastic process. A saliency map is used to represent a landscape upon which non-local transition probabilities generate a constrained random walk.

More[16] recently, Zanca et al. proposed an approach that is inspired by physics to model the process of visual attention as a continuous dynamic phenomenon [90, 92]. The focus of attention is modeled as a point-like particle gravitationally attracted by virtual masses originated from details and motion in the visual scene. Masses due to details are determined by the magnitude of the gradient of the brightness, while masses due to motion are proportional to the norm of the optical flow. This framework can be applied to both images and videos, as long as one considers a static image as a video whose frames are repeated at each time step. Moreover, the model proposed in [92] also implements the inhibition of return mechanism, by decreasing the saliency of a given area of the retina that has already been explored in the previous moments. Unlike the other described approaches, the prediction of the focus does not rely on a centralized saliency map, but it acts directly on early representations of basic features organized in spatial maps. Besides the advantage in real-time applications, these models make it possible to characterize patterns of eye movements (such as *fixations*, *saccades*, and *smooth pursuit*) and, despite their simplicity, they reach the state of the art in scanpath prediction and proved to predict shifts in visual attention better than the classic winner-take-all [91].[17] However, when looking at these gravitational models from the biological and computational

---

[16] Virtual masses and gravitational models

[17] Gravitational models are non-local in space and, therefore, are not biologically plausible

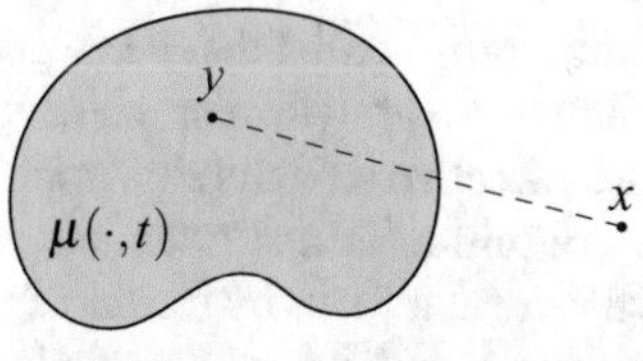

$$u^0(x,t) = \frac{1}{2\pi} \int_{\mathbb{R}^2} \log \frac{1}{\|x-y\|} \mu(y,t)\,dy$$

Fig. 2.1 Gravitational model of focus of attention. The cognitive process of focus of attention follows the model of mass attraction, which turn out to be the translation of spatiotemporal visual details. Here, we have displayed a general density distribution and the corresponding 2-D potential

perspective, one promptly realizes that finding the focus of attention at a certain time does require the access to all the visual information of the retina to sum up the attraction arising from any virtual mass. Basically, those models are not local in space.

While current computational models keep improving their predictive ability thanks to the increasing availability of data, they are still far away from the effectiveness and efficiency exhibited by foveated animals. An in-depth investigation on biologically plausible computational models of focus of attention that exhibit spatiotemporal locality is very important also for computer vision, where one relies on parallel and distributed implementations. The ideas given in [30], properly re-elaborated, are herein proposed mostly because of their effectiveness and biological plausibility.[18] A computational model is proposed where attention emerges as a wave propagation process originated by visual stimuli corresponding to details and motion information. The resulting field obeys the principle of *inhibition of return*, so as not to get stuck in potential holes, and extends previous studies in [92] with the main objective of providing spatiotemporal locality. In particular, the idea of modeling the focus of attention by a gravitational process finds its evolution in the corresponding local model based on Poisson equation on the corresponding potential. Interestingly, Newtonian gravity yields an instantaneous propagation of signals, so as a sudden change in the mass density of a given pixel immediately affects the focus of attention, regardless of its location on the retina. These studies are driven by the principle that there are in fact sources which drive attention (e.g., masses in a gravitational field) (Fig. 2.1).

Here, we discuss a paradigm shift in the computation of the attraction model proposed in [92], which is inspired by the classic link between global gravitational or electrostatic forces and the associated Poisson equation on the corresponding potential that can be regarded as a spatially local computational model. While the link is intriguing, modeling the focus of attention by the force emerging from the static nature of the Poisson potential does not give rise to a truly local computational process, since one needs solving the Poisson equation for each frame. This means

---

[18] Visual stimuli corresponding to spatial details and motion turn out to act as sources for the focus of attention field and originate wave propagation which, unlike for the gravitational field, has a finite propagation velocity

that such a static model is still missing the temporal propagation that takes place in peripheral vision mechanisms. We show that the temporal dynamics which arises from diffusion and wave-based mechanisms are effective to naturally implement local computation in both time and space. The intuition is that attention is also driven by virtual masses that are far away from the current focus by means of wave-based and/or diffusion propagation. We discuss the two different mechanisms of propagation and prove their reduction to gravitational forces as the velocity of the propagation goes to infinity. The experimental results confirm that the information coming from virtual masses is properly transmitted on the entire retina. In the case of video signals, our experimental analysis on scanpaths leads to state-of-the-art results which can clearly be interpreted when considering the reduction to the gravitational model [92] for infinite propagation velocity. The bottom-up is that we can reach state-of-the-art results by a computational model which is truly local in space and time. As a consequence, the proposed model is very well suited for single instruction multiple data (SIMD) specialized implementations. Moreover, the proposed theory on focus of attention sheds light on the way it takes place in biological processes [72].[19]

According to the gravitational model of [92], the trajectory of the focus of attention $t \in [0, T] \mapsto a(t) \in \mathbb{R}^2$, starting at $a(0) = a_0$ with velocity $\dot{a}(0) = a_1$, is the solution of the following Cauchy problem:

$$\begin{cases} \ddot{a}(t) + \varpi \dot{a}(t) + \nabla u^0(a(t), t) = 0; \\ a(0) = a_0; \\ \dot{a}(0) = a_1, \end{cases} \tag{2.1}$$

where $\varpi > 0$ and the scalar function $u^0 \colon \mathbb{R}^2 \times [0, T] \to$ is defined as follows:

$$u^0(x, t) := \frac{1}{2\pi} \int_{\mathbb{R}^2} \log \frac{1}{\|x - y\|} \mu(y, t) \, dy. \tag{2.2}$$

Here, $\|x - y\|$ is the Euclidean norm in $\mathbb{R}^2$ and $\mu \colon \Omega \subset \mathbb{R}^2 \times [0, T] \to [0, +\infty)$ is the mass distribution at a certain temporal instant that is present on the retina and that is responsible of "attracting" the focus of attention. Such mass distribution involves details and motion, and formally, it is determined by

$$\mu(x, t) = \mu_1(x, t) (1 - I(x, t)) + \mu_2(x, t). \tag{2.3}$$

In[20] particular, $\mu_1 = \alpha_1 \|\nabla b\|$, where $b \colon \mathbb{R} \times [0, T] \to \mathbb{R}$ is the brightness, while $\mu_2 = \alpha_2 \|\partial_t b\|$, while α_1 and α_2 are positive parameters. The term $I(x, t)$ implements the inhibition of return mechanism, and it satisfies

---

[19] FOA driven by gravitation that is characterized by the corresponding potential φ^0

[20] The role of the inhibition of return function I which drives the change of the virtual mass

$$I_t + \beta I = \beta \exp(-\|x - a(t)\|^2 / 2\sigma^2), \tag{2.4}$$

with $0 < \beta < 1$ (I_t is the time derivative of I). We can promptly see that this model for I nicely implements the mechanism for avoiding to get stuck in the focus of attention. If x approaches $a(t)$, then the system dynamics of I_t leads to $I_t \simeq 1$, with a velocity which depends on β. As $I_t \simeq 1$, from Eq. (2.4) we see that the component of the virtual mass coming from spatial details disappears, that is, favoring the escape from the focus of attention. Because of the need to know the distance from the focus of attention from Eq. (2.2), we can see that the model of [92] is not local in space.

Notice that the potential u^0 satisfies the Poisson equation on $\mathbb{R}^2$:

$$-\nabla^2 u = \mu, \tag{2.5}$$

where ∇^2 is the *Laplacian* in two dimensions. Such result, which is the two-dimensional analogue of the Poisson equation for the classical gravitational potential, can be checked by direct calculation. Because the mass density μ is time dependent, and its temporal dynamics is synced with the temporal variations of the video, Eq. (2.5) should in principle be solved for any t. We will discuss how the values of the potential in a spatial neighbor of (x, t) are exploited to estimate the values of u at $(x, t + dt)$ by interpreting Eq. (2.5) as an elliptic limit of a parabolic or hyperbolic equation instead.

Evaluating the gravitational force acting on the FOA starting with the Poisson equation requires to compute the potential $u(x, t)$ due to the virtual masses at each frame, thus ignoring any temporal relation. This key remark also underlines the strong limitation of the solution proposed in [92] (and also in [90]), where the gravitational force is re-computed at each frame from scratch. The main idea behind the reformulation presented here is that since we expect that small temporal changes in the source μ cause small changes in the solution u, then it is natural to model the potential u by dynamical equations which prescribe, for each spatial point x, how the solution must be updated depending on the spatial neighborhood of x at time $t - dt$.

We[21] can introduce an explicit temporal dynamics in Eq. (2.5) by introducing the two following "regularizations"

$$
\begin{aligned}
H : &\begin{cases} c^{-1} u_t = \nabla^2 u + \mu & \text{in } \mathbb{R}^2 \times (0, +\infty); \\ u(x, 0) = 0, & \text{in } \mathbb{R}^2 \times \{t = 0\}, \end{cases} \\
W : &\begin{cases} c^{-2} u_{tt} = \nabla^2 u + \mu & \text{in } \mathbb{R}^2 \times (0, +\infty); \\ u(x, 0) = 0, \quad u_t(x, 0) = 0 & \text{in } \mathbb{R}^2 \times \{t = 0\}, \end{cases}
\end{aligned}
\tag{2.6}
$$

where $c > 0$ and u_t (u_{tt}) is the first (second) time derivative of u. Problem H is a Cauchy problem for the heat equation with source $\mu(x, t)$, whereas problem W is a Cauchy problem for a wave equation. The term c in H represents the diffusivity constant, whereas the constant c in problem W can be regarded as the wave propagation

---

[21] Biologically plausible wave propagation of FOA

velocity. The reason why we can consider problem H and W as *temporal regularizations* of Eq. (2.5) is due to a fundamental link [30] between the gradients ∇u_H and ∇u_W of the solutions u_H and u_W to problems H and W in Eq. (2.6) with u^0, which is the solution, described in Eq. (2.2). Those gradients for H and W both converge to ∇u^0 as $c \rightarrow +\infty$. The interpretation of this result is actually quite straightforward. For problem H, it means that the solution of the heat equation in a substances with high diffusivity c instantly converges to its stationary value which is given by Poisson equation (2.5). For problem w, the above convergence statement turns out to be the two-dimensional analogue of the infinite-speed-of-light limit in electrodynamics and in particular it expresses the fact that the retarded potential (see [52]), which in three spatial dimensions are the solutions of problem W, converges to the electrostatic potential as the speed of propagation of the wave goes to infinity ($c \rightarrow +\infty$).[22]

Although both temporal regularizations H and W achieve the goal of transforming the Poisson equation into an initial value problem in time from which all subsequent states can be evolved from, a different nature of the two PDE determines, for finite c, qualitative differences in the FOA trajectories computed using Eq. (2.1). In the remainder, we then consider the following generalized version of Eq. (2.6)

$$\begin{cases} \gamma u_{tt}(x,t) + \lambda u_t(x,t) = c^2 \nabla^2 u(x,t) + \mu(x,t) & \text{in } \mathbb{R}^2 \times (0,+\infty); \\ u(x,0) = 0, \quad u_t(x,0) = 0 & \text{in } \mathbb{R}^2 \times \{t = 0\}, \end{cases} \quad (2.7)$$

where $\lambda \geq 0$ is the drag coefficient and $\gamma \geq 0$. Such equation in one spatial dimension (and without the source term μ) is known as the telegraph equation (see [29]). More generally, it describes the propagation of a damped wave. The pure diffusion case H corresponds to $\gamma = 0$, with a diffusion coefficient equal to $\alpha = c^2/\lambda$ and a source term of $\mu(x,t)/\lambda$. With $\gamma = 1$ and $\lambda = 0$, we obtain the pure wave equation W instead. The FOA model proposed here is based on Eq. (2.7) along with the inhibition of return equation expressed by (2.4).

Clearly, Eq. (2.7) is local in both space and time, which is a fundamental ingredient of biological plausibility. In addition, they are very well suited for SIMD hardware implementations. At a first sight, Eq. (2.4) does not possess spatial locality. While this holds true in any computer-based retina, in nature, moving eyes rely on the principle that you can simply precompute $\exp(-\|x - a(t)\|^2/2\sigma^2)$ by an appropriate foveal structure. Interestingly, the implementation of moving eyes is the subject of remarkable interest in robotics for different reasons (see, e.g., [60]).[23]

> Driving the focus of attention is definitely a crucial issue. We conjecture that this driving process must undergo a developmental process, where we begin

---

[22] It is worth mentioning that while this kind of regularization is well known in three dimensions, the same properly has not been formally stated in two dimensions. A formal proof of the property in the case of two dimensions is given in [30].

[23] Question no. 8: What drives focus of attention?

with details and optical flow and we carry on with the fundamental feedback from the environment which is clearly defined by the specific purpose of the agent.

2.6 The Virtuous Loop of Focus of Attention

In haplorhine primates, the process of learning to see is interwound with that of focusing attention. We can access to their strict conjugation by cognitive statements but, as it will become more clear in the next chapter, we can also give such a tight connection an in-depth mathematical formulation.

During[24] early cognitive stages, in newborns attention mechanisms are mostly driven by the presence of details and movements. However, this is not the whole story. Humans are endowed with an exceptional ability for detecting faces, and already shortly after birth, they preferentially orient to faces. Simple tests of preference have been carried out which provide evidence on the fact that infants as young as newborns prefer faces and face-like stimuli over distractors [63]. However, the issue is quite controversial and there is currently no agreement as to how specific or general are the mechanisms underlying newborns' face preferences [31]. The neural substrates underlying this early preference are also quite an important subject of investigation. In [23], the authors point out that measured EEG responses in 1- to 4-day-old infants and discovered reliable frequency-tagged responses when presenting faces. Upright face-like stimuli elicited a significantly stronger frequency-tagged response than inverted face-like controls in a large set of electrodes. Overall, their results suggest that a sort of cortical route specialization is already functional at birth for face perception. This specialization makes sense. In other species of animals, the degree of visual skills that are exhibited at the birth is remarkably higher than in humans, which is clearly connected with very different environmental constraints. From the viewpoint of information theory, the degree of visual skills that are exhibited at the birth is related to the genetic heritage. There is an interesting trade-off in nature between the transmitted visual skills and the transmitted mechanisms for their learning. Focus of attention does not escape this general rule, but it seems to be strongly biased toward the learning component, so as early behavior is quite elementary. This is the reason why we stressed the analysis of a model based on the magnitude of the spatiotemporal gradient of the brightness.

In children, the mechanisms that drive the focus of attention are strongly connected with the developmental stages. As already stated, at early stages of life, they only focus attention on details and movements with some preference for visual patterns like faces. As time goes by, visual features conquer a semantic value and, consequently, the focus of attention is gradually driven by specific intentions and

---

[24] Where do you focus attention in the early stages of life?

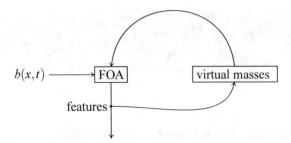

Fig. 2.2 Virtuous loop of focus of attention. After a phase in which FOA is driven by visual details that are regarded as virtual masses, as time goes by, the development of visual features results in new virtual masses that also provide focus of attention by means of the same attraction mechanisms

corresponding plans. Of course, this is only possible after having acquired some preliminary capability of recognizing objects. Interestingly, as the forward process that facilitates high-level cognitive tasks from the focus of attention becomes effective a corresponding backward process begins the improvement of the focus of attention. A reinforcement loop is generated which is finalized to optimize the final purpose of the agent in its own learning environment as it is shown in Fig. 2.2.

The[25] driving mechanisms of the FOA, as it emerges from the virtual masses connected with $\nabla b = [\nabla_x b, \partial_t b]$, are clearly primitive! In many species of animals—surely in humans—the driving process behind the FOA is definitely more complex. We use to look at "something interesting" and do not just focus on details. Not all details are in fact interesting, and it looks like we early end up into circular issues when trying to establish what "interesting" means. When restricting our scope to vision, we can decouple the process of eye movements with the intention which is purposely driving our attention to specific tasks. In order to exhibit such a skill, we already need to possess the competence of visual perception we are interested in. Hence, it looks like we cannot neglect a developmental structure in the emergence of FOA and, correspondingly, in the acquisition of visual skills. Hence, the early mechanisms of FOA could be actually based on the field generated by sources corresponding to visual details, defined by ∇b. Is there anything better? Can a visual agent refine the skills gained using visual details as time goes by? As will see in detail in the next chapter, we are primarily interested in tracking any visual structure with some meaning for the agent more than single pixels. The immediate consequence is that the virtual masses used so far should be updated to something meaningful that goes beyond the brightness of the pixels! A convolutional net, with its visual features, clearly offers the possible solution: Instead of simply considering virtual masses based on ∇b, we can consider virtual masses associated with the visual features themselves. This gives rise to a virtuous cycle: From initial acquisition of FOA based on ∇b, we learn visual features, which in turn become virtual masses to be used for a further improvement of FOA. This cycle reinforces the mutual acquisition of FOA and visual

---

[25] The FOA loop and the *duality principle*

features, which become somehow conjugate. At any stage of development, the agent drives attention on the basis of his current visual features, which are consistent with FOA movement. This interesting cyclic behavior can be interpreted in terms of the *duality principle* which leads to regard motion and features as two sides of the same medal. Interestingly, while motion invariance is a mechanism for feature learning based on the external visual source on the opposite, FOA is an internal mechanism for generating the agent's velocity from features.

Finally,[26] we need to say something more about the way the FOA loop is activated. In the simplest case, it can be stimulated simply by the brightness, along with its spatiotemporal changes that give rise to the virtual masses which draw attention. Of course, the presence of richer features at the birth produces different virtual masses capable of driving more sophisticated focusing mechanisms. No matter what the specific choice is, any visual agent which misses the activation of FOA is basically a frog: The agent cannot incorporate environmental motion invariance because of the *drifting problem*. The introduction of mechanisms for avoiding to get stuck focusing in single points of the retina (e.g., see the inhibition of return) plays a fundamental role in the introduction of saccadic movements. Overall, they play a crucial role in the computational mechanisms behind the development of motion invariance.

---

[26] FOA is the first field which needs to be activated

Chapter 3
Principles of Motion Invariance

In science there is and will remain a Platonic element which could not be taken away without ruining it. Among the infinite diversity of singular phenomena science can only look for invariants.

Jacques Monod, 1971

Keywords Motion invariance · Optical flow · Affordance

3.1 Introduction

The strong assumption used throughout this book is that any information-based interpretation of visual perception only relies on motion. Unless we are only interested in pairing visual and linguistic skills in predefined application domains, *all you need is motion invariance*, which we claim is at the origin of the development of visual skills in all living animals. Basically, this claim is directly facing the supervised learning approach that has already successfully shown the capabilities of deep convolutional networks. Their spectacular achievements clearly indicate the fundamental role of deep architectures in the learning of very complex functions by means of a number of parameters that could not be even considered ten years ago. However, while the emergence of such a power of deep convolution nets is now widely recognized, the associated supervised learning protocol, which has dominated the computer vision applications in the last few years, is currently the subject of heated discussions. The extraordinary visual abilities of the eagle and other animals seem to suggest that the fundamental processes of vision obey visual interaction protocols that go beyond supervised learning. In this book, we address primarily the question on whether we can replace the fundamental role of huge labeled databases with the simple "life in a visual environment" by means of an appropriate interpretation of the information coming from motion.[1]

---

[1] Motion is all what you need: the two principles of motion invariance.

A. Betti et al., *Deep Learning to See*, SpringerBriefs in Computer Science, https://doi.org/10.1007/978-3-030-90987-1_3

A major claim here is that motion is all what you need for extracting information from a visual source. Motion is what offers us an object in all its poses. Classic translation, scale, and rotation invariances can clearly be gained by appropriate movements of a given object. However, the experimentation of visual interaction due to motion goes well beyond and includes the object deformation, as well as its obstruction. Only small portions can be enough for object detection, even in the presence of environmental noise. How can motion be exploited? In this chapter, we establish two fundamental principles of visual perception that shed light on the acquisition of the identity and on the abstract notion of object as perceived by humans.[2] The *first principle of visual perception* involves consistency issues, namely the preservation of material points during motion. Depending on the pose, some of those points are projected onto the retina, and others are hidden. Basically, the material points of an object are subject to *motion invariance of the corresponding pixels on the retina*. A moving object clearly does not change its identity, and therefore, imposing motion invariance conveys crucial information on its recognition. Interestingly, more than the recognition of an object category, this leads to the discovering of its identity. Motion information does not only confer object identity, but also its affordance, its function in real life. Affordance makes sense for a species of animal, where specific actions take place. A chair, for example, has the affordance of seating a human being, but can have many other potential uses. The *second principle of visual perception* is about its affordance as transmitted by coupled objects—typically humans. The principle states that the affordance is invariant under the coupled object movement. Hence, a chair gains the seating affordance independently of the movement of the person who's seating (coupled object).[3]

As already seen in the previous chapter concerning FOA, there are a number of reasons for adopting a field theory of vision, where velocities, brightness, and features turn out to be associated with a given pixel. This makes it possible to establish motion invariance principles in all their facets, without getting stuck in the chicken–egg dilemma mentioned in Sect. 1.4. This is very much in line with the very well-established literature in the field of optical flow, where the emphasis is on making sense of the perceived velocity on the retina. In many real-world cases, the velocity of material points projected onto the retina differs from the perceived velocity, which is referred to as the optical flow. Interestingly, we shall explore richer interpretations of the optical flow by the introduction of velocities that are referred to specific features. Overall, features and velocities are vision fields. This chapter is organized as follows. In the next section, we introduce the idea of learning in spatiotemporal environments, while in Sect. 3.3 we discuss the fundamental distinction between object identity and its affordance. In Sect. 3.4, we introduce the classic notion of optical flow while discussing how a map can be established from the visible material points to the retina. In Sects. 3.5 and 3.6, we present the two fundamental principles of motion invariance connected with object identity and affordance, respectively. Finally, in

---

[2] I and II Principles of visual perception: object identity and affordance.

[3] A field theory for vision.

Sect. 3.7 we extend motion invariance as presented by the two previous principles to the general case of features.

3.2 Computational Models in Spatiotemporal Environments

For many years, scientists and engineers have proposed solutions to extract visual features that are mostly based on clever heuristics (see, e.g., [67]), but more recent results indicate that most remarkable achievements have come from deep learning approaches that have significantly exploited the accumulation of huge visual databases labeled by crowdsourcing [59, 80].

However, in spite of successful results in specific applications, the massive adoption of supervised learning leads to face artificial problems that, from a pure computational point of view, are likely to be significantly more complex than natural visual tasks that are daily faced by animals. In humans, the emergence of cognition from visual environments is interwound with language. However, when observing the spectacular skills of the eagle that catches the prey, one promptly realizes that for an in-depth understanding of vision, one should begin with a neat separation from language![4]

What are the major differences between human and machine learning in vision? Can we really state that the "gradual" process of human learning is somewhat related to the "gradual" weight updating of neural networks? A closer look at the mechanisms that drive learning in vision tasks suggests that nowadays models of machine learning mostly disregard the fundamental role of "time," which should not be confused with the iteration steps that mark the weight update. First, notice that in nature learning to see takes place in a context where the classic partition into learning and test environment is arguable. On the other hand, this can be traced back to early ideas on statistical machine learning and pattern recognition, which are dominated by the principles of statistics. The gathering of the training, validation, and test data leads to discover algorithms that are expected to find regularities in big data collections. This is good, yet it enables artificial processes whose underlying communication protocol might not be adequate in many real-world problems.[5] In the extreme case of batch mode learning, the protocol assumes that the agent possesses information on its life before coming to life. Apparently, this does not surprise computer vision researchers, whereas it sounds odd for the layman, whose viewpoint should not be neglected, since we might be trapped into an artificial world created when no alternative choice was on the horizon. The adoption of mini-batches and even the extreme solution of online stochastic gradient learning are still missing a truly incorporation of time. Basically, they pass through the whole training data many times, a process which is still far

---

[4] What is the role of time?

[5] Batch mode learning: In this extreme case, the agent is expected to know information on its visual interaction over all its life before coming to life!

apart from natural visual processes, where the causal structure of time dictates the gradual exposition to video sources. There is a notion of lifelong learning that is not captured in nowadays computational schemes, since we are ignoring the role of time which imposes causality.

Interestingly, when we start exploring the alternatives to huge collection of labeled images, we are immediately faced with a fundamental choice, which arises when considering their replacement with video collections. What about their effectiveness? Suppose you want to prepare a video collection to be used for learning the market segment associated with cars (luxury vehicles, sport cars, SUVs/off-road vehicles, etc.). It could be the case that a relatively small image database composed of a few thousands of labeled examples is sufficient to learn the concept. On the other hand, in a related video setting with the same space constraint, this corresponds with a few minutes of video, a time interval in which it is unreasonable to cover the variety of car features that can be extracted from 10,000 images! Basically, there will be a lot of nearly repetitions of frames which support scarce information with respect to the abrupt change from picture to picture. This is what motivates a true paradigm shift in the formulation of learning theories for vision. In nature, it is unlikely to expect the emergence of vision from the accumulation of video.[6] Hence, could not machines do the same? A new communication protocol can be defined where the agent is simply expected to learn by processing the video as time goes by without its recording. Interestingly, this might open great opportunities to all research centers to compete in another battlefield.

> The bottom line is that while we struggle for the acquisition of huge labeled databases, the actual incorporation of time might lead to a paradigm shift in the process of feature extraction. We promote the study of the agent life based on the ordinary notion of time, which emerges in all its facets. The incorporation of motion invariance might be the key for overcoming the artificial protocol of supervised learning. We claim that such an invariance is in fact the only one that we need.

This comparison between animals and computers leads to figure out what human life could have been in a world of visual information with shuffled frames. A related issue has been faced in [88] for the acquisition of visual skills in chicks. It is pointed out that "when newborn chicks were raised with virtual objects that moved smoothly over time, the chicks developed accurate color recognition, shape recognition, and color–shape binding abilities." Interestingly, the authors notice that in contrast, "when newborn chicks were raised with virtual objects that moved non-smoothly over time, the chicks' object recognition abilities were severely impaired." When exposed to a video composed of independent frames taken from a visual database, like ImageNet,

---

[6] Question no 2: Animals gradually conquer visual skills in their own environments. Can computer do the same?

that are presented at classic cinema frame rate of 24 fps, humans seem to experiment related difficulties in such a non-smooth visual presentation.[7]

Interestingly, it turns out that our visual skills completely collapse in a task that is successfully faced in computer vision! As a consequence, one might start formulating conjectures on the inherent difficulty of artificial versus natural visual tasks. The remarkably different performances of humans and machines have stimulated the curiosity of many researchers in the field. Of course, you can start noticing that in a world of shuffled frames, a video requires order of magnitude more information for its compressed storing than the corresponding temporally coherent visual stream. This is a serious warning that is typically neglected in computer vision, since it suggests that any recognition process is likely to be more difficult when shuffling frames. One needs to extract information by only exploiting spatial regularities in the retina, while disregarding the spatiotemporal structure that is offered by nature. The removal of the thread that nature used to sew the visual frames might prevent us from the construction of a good theory of vision. Basically, we need to go beyond the current scientific peaceful interlude and abandon the safe model of restricting computer vision to the processing of images. Working with video was discouraged at the dawn of computer vision because of the heavy computational resources that it requires, but the time has come to reconsider significantly this implicit choice. Not only humans and animals cannot see in a world of shuffled frames, but we also conjecture that they could not learn to see in such an environment. Shuffling visual frames is the implicit assumption of most of nowadays computer vision approaches that, as stated in the previous section, corresponds with neglecting the role of time in the discovering of visual regularities. No matter what computational scheme we conceive, the presentation of frames where we have removed the temporal structure exposes visual agents to a problem where a remarkable amount of information is delivered at any presentation of new examples. When going back to the previous discussion on time, one clearly sees that the natural environmental flow must be somehow synchronized with the agent computational capability. The need for this synchronization is in fact one of the reasons for focusing attention at specific positions in the retina, which confers the agent also the gradual capability of extracting information at pixel label. Moreover, as already pointed out, we need to abandon the idea of recording a database for statistical assessment. There is nothing better than human evaluation in perceptual tasks, which could stimulate new ways of measuring the scientific progress of computer vision.

The reason for formulating a theory of learning on video instead of on images is not only rooted in the curiosity of grasping the computational mechanisms that take place in nature.[8] A major claim in this book is that those computational mechanisms are also fundamental in most of computer vision tasks.

---

[7] Can animals see in a world of shuffled frames?

[8] Question no. 3 on shuffling of video frames; does it increase the complexity of learning to see?

It looks like that, while ignoring the crucial role of temporal coherence, the formulation of most of nowadays current computer vision tasks leads us to tackle problems that are remarkably more difficult than those nature has prepared for us!

3.3 Object Identity and Affordance

Human visual skills are somewhat astonishing. We get in touch with the spectacular capabilities of interpreting visual scene whenever we open the challenge of describing the underlying computational process. From one side, humans can recognize the objects' identity regardless of the different poses in which they come. This is quite surprising! Objects are available in different positions; at different scales and rotations, they can be deformable and obstructed. Only small portions can be enough for their detection, even in the presence of environmental noise.

I recognize my plaid blanket that I put on the sofa from the others in my house. It can be folded more or less well, creased, stretched out, piled up somewhere, but it is still here. If I squeeze a Perrier plastic bottle with my hands and close it with its cap so that it maintains the flattened shape, I continue to recognize it without difficulty. If I accidentally break my favorite cup, I can still recognize it from many of its pieces, maybe not all of them, a bit like when I see it despite being obstructed by other objects. Children recognize their teddy bear from different angles. As they bite it, not only does the bear deform as a result, but they experience a very peculiar sight of it. Yet, they perceive it consistently.[9]

However, you cannot really hope to recognize your bear by a front picture, if the only thing that distinguishes it from another is that it is slightly torn behind the head. It is that torn part that gives it its identity. It looks like the object motion is always behind the different poses of the object. There is nothing else than motion which concurs to present the different ways we perceive an object, with its own identity. Hence, one might wonder whether the object identity comes from the massive number of poses provided by its motion. The corresponding learning process should develop such an invariance by neural-based maps. Deep learning has already been proven to be very effective to replicate those visual capabilities, provided that we rely on tons of supervisions. Interestingly, it could be case that those explicit linguistic-based supervisions can simply be offered by motion. When considering the virtually infinite availability of visual information offered by motion, the supervised learning approach may have hard time to reach the human capabilities to define the identity of objects.

Any learning process that relies on the motion of the given object can only aspire to discover its identity. The motion invariance process is in fact centered around the

---

[9] Object identity: Does it emerge from the massive number of poses provided by its motion?

Fig. 3.1 On the left side, we can see the same chair in different positions, scales, and rotation. On the right side, different types of chairs are shown as an example of gaining the abstract notion of chair

object itself, and as such, it does reveal its own features in all possible expositions that are gained during motion. However, as shown in Fig. 3.1, in addition to the capability of identifying objects (left side), humans clearly gain abstraction (right side).[10] Humans, and likely most animals, also conquer a truly different understanding of visual scene that goes beyond the conceptualization with single object identities. In early sixties, James J. Gibson coined the *affordance* in [37], even though a more refined analysis came later in [38]. In his own words:

> *The affordances of the environment are what it offers the animal, what it provides or furnishes, either for good or ill. The verb to* afford *is found in the dictionary, the noun affordance is not. I have made it up. I mean by it something that refers to both the environment and the animal in a way that no existing term does. It implies the complementarity of the animal and the environment.*

Affordance makes sense for any species of animal, where specific actions take place. A chair, for example, has the affordance of seating a human being, but can have many other potential uses. We can climb on it to reach a certain object, or we can use it to defend ourselves from someone armed with a knife. It was also early clear that no all objects are significantly characterized by their affordance. In [38], the distinction between *attached and detached objects* clearly indicates that some can be neither moved nor easily broken. While our home can be given an affordance, it is attached to the earth and cannot be grasped. Interestingly, while a small cube is graspable, huge cubes are not. However, a huge cube can still be graspable by an appropriate "handle." Basically, one can think of object perception in terms of its properties and qualities. However, this is strictly connected to the object identity, whereas most common and useful ways of gaining the notion of objects likely come from their affordances. The object affordance is what humans—and animals—normally pay attention to.

---

[10] Object affordance.

As infants begin their environmental experiences, they likely look at object mean-ing more then at the surface, at the color or at its form. This interpretation is kept during our ordinary environmental interactions.[11] The notion of object affordance somewhat challenges the apparently indisputable philosophical muddle of assuming fixed classes of objects. This is a slippery issue! To what extent can we regard object recognition as a classification problem? Clearly, there is no need to classify objects in order to understand what they are for. Hence, we could consider the advantages of moving from nowadays dominating computer vision approach to a truly ecological viewpoint that suggests experimenting niches-based set of well-defined, typically limited object affordances.

3.4   From Material Points to Pixels

First, we begin discussing the classic notion of optical flow, which is in fact of central importance in computer vision.[12] It is a slippery concept, and we shall see that its in-depth analysis is very important for interesting developments on more general tracking issues.[13] In computer vision, one typically considers the case of moving objects acquired by a fixed camera as well as the case of egocentric vision that entails analyzing images and videos captured by a wearable camera. However, because of eye movements, foveated animals experiment the presence of optical flow even in the case in which there is neither the movement of the animal nor of the visual environment. Hence, one can always consider to deal with the general case in which the visual information is acquired in the reference of the retina, where we are always in front of optical flow.

The object movements in the real world can be naturally framed in a three-dimensional space (3-D), and as such, the velocity of any single material point turns out to be a 3-D vector. Animals and computers can only perceive the movement as a 2-D projection of material points onto the correspondent pixels of the retina. Clearly, finding such a correspondence is not an easy problem. First, some material points are not visible. When they are visible, the map that transforms material points to pixels is not necessarily unique, especially for visually uniform regions. A trivial case in which ill-position arises is simply when the camera frames an entire white picture. There is no way to tell if something is moving; it could be either a wall or a big moving white object. While one can always exclude this case, the underlying ambiguity still remain in ordinary visual environments.[14]

An enlightening approach to the problem of estimating the optical flow of the pix-els corresponding to material points of a moving object is that of imposing the princi-ple of *brightness invariance*. Let $b(x, t)$ be the brightness at $(x, t) \in \Omega \times (0, T) =:$

---

[11] Do we really need to classify objects?

[12] What is the optical flow?

[13] Foveated animals always experiment motion!

[14] The principle of brightness invariance.

Γ. In the big picture of this book, T denotes the duration of video segments delimited by saccadic movements. The principle consists of stating that the moving point associated with the trajectory $x(t)$ does not change its brightness, that is, $b(x(t), t) = c$, being c constant over time. Hence, this can be restated by

$$\frac{db(x(t), t)}{dt} = \nabla b \cdot v + b_t = 0, \quad \forall t \in (0, T). \tag{3.1}$$

This is a condition on the velocity $v = \dot{x}$ that can admit infinite solutions since it is a scalar equation with two unknowns. An enlightening approach to the problem of estimating the optical flow of the pixels corresponding to material points of a moving object was given by Horn and Schunck in a seminal paper published at the beginning of the eighties [48]. They adopted a regularization principle which turns out to have a global effect on the guess of the optical flow. Basically, they proposed to determine the optical flow v as the minimization of

$$E(v_1, v_2) = \int_\Omega \left(\nabla v_1\right)^2 + \left(\nabla v_2\right)^2, \tag{3.2}$$

subject to the constraint (3.1). One can promptly realize that while this is a well-posed formulation, it does not take into account the change of brightness of moving material points when lighting conditions change significantly. If the object moves from a luminous to a dark area, then the condition (3.1) does not really reflect reality and the corresponding estimation of the velocity becomes quite poor.[15]

Notice that the brightness might not necessarily be the ideal signal to track. Since the brightness can be expressed in terms of the red R, green G, and B components as $b = R + G + B$, one could think of tracking single color components of the video signal by using the same invariance principle stated by Eq. (3.1). It could in fact be the case that one or more of the components R, G, B are more invariant during the motion of the corresponding material point. In that case, in general, each color can be associated with a corresponding velocity v_R, v_G, v_B that might be different. In so doing, instead of tracking the brightness, one can track the single colors. On the opposite, the simultaneous track of all the colors, with the same velocity $v = v_R = v_G = v_B$, yields

$$v \cdot \nabla \begin{pmatrix} R \\ G \\ B \end{pmatrix} + \frac{\partial}{\partial t} \begin{pmatrix} R \\ G \\ B \end{pmatrix} = 0, \tag{3.3}$$

where $v \cdot \nabla(R, G, B) := (v \cdot \nabla R, v \cdot \nabla G, v \cdot \nabla B)$ and $\nabla(R, G, B)$ is generally nearly singular. However, it is worth mentioning that the simultaneous tracking of different channels might contribute to a better positioning of the problem. One can think of the color component as features that unlike classic convolutional spatial

---

[15] Color tracking.

features are temporal features. We can in fact regard the color components as the outputs of three convolutional filters in the temporal domain.

Interestingly, as it will be discussed in the following, humans and likely many species of animals very well deal with this problem when they are involved in tracking tasks. An in-depth analysis on this issue leads to a novel of velocity which is interwound with the notion of visual feature. Before facing this major topics, it is convenient to better explore the meaning of the optical flow and grasp the limitations behind the principle of brightness invariance.
Problems with the principle of brightness invariance

When the brightness of the background changes, an illusion phenomenon can arise that do compromise the correct estimation of the actual kinematic velocity. Let us consider an artificial example to illustrate this issue. In particular, suppose that

$$b(x, t) = tx \cdot \hat{e}_1 \tag{3.4}$$

where $\hat{e}_1 = (1, 0)'$, and then a solution of the brightness invariance condition is $v(x, t) = -(x \cdot \hat{e}_1/t, 0)'$. Notice that here the choice $v \cdot \hat{e}_2 = 0$ has been done quite arbitrarily since any value of the projection of the velocity along $\hat{e}_2$ would satisfy the brightness invariance condition. Anyway, this example shows how a fixed image with changing lighting over time can generate an illusion of motion.

3.5   The Principle of Material Point Invariance

A cat that is chasing a mouse continues its chase effectively even if the mouse passes through areas of spatially or temporally variable brightness like the one considered in Eq. (3.4). Likewise, a flickering light does not significantly help the mouse escape. How can this be possible? The cat is not only tracking single material points, but he has clearly got the capability of perceiving the overall picture of the mouse, as well as many of his distinguishing features. Hence, tracking in nature involves directly objects and their features more than single material points.[16]

---

[16] The cat chasing the mouse: What is the velocity of the mouse estimated by the cat?

As the cat chases the mouse, it looks like he is given the task to estimate the mouse velocity. This does require a deeper analysis. Are we talking of the velocity of the barycenter of the mouse? How can the cat determine the barycenter? Again, the pixel-wise computation that has been invoked in Sect. 2.2 comes to help. We can think of convolutional-like features on (x, t) that somehow characterize the way the mouse is visually interpreted by the cat, who is tracking those features. We can associate different velocities at (x, t) depending on the feature we are referring to. As the feature conquers an interpretable semantics, the velocities associated with each pixel offer a novel scenario for the cat, who does not restrict to the flow of the velocity associated with the material points.[17] In Sect. 3.3, we introduced the fundamental distinction between object identity and affordance. We begin addressing all issues involving object identity, since it covers motion invariance of the object features. The process of recognizing the object identity is based on the following *material point invariance principle* (MPI) (I Principle), which blesses the pairing of any feature of a given object, including the brightness, with its own velocity.

First Principle of Perceptual Vision: Material Point Invariance

Let us consider the motion of an object $\mathcal{O}$ over Γ. As already mentioned, we are considering burst of movements that can be delimited by saccadic jumps of duration T at most. The reason for this limited duration has already been discussed when we introduced the problem of the concept drift.[18] Now, let us denote by $P(t) \in \mathcal{O}$ one of its material points of object $\mathcal{O}$ at $t < T$. Let $x(t) = \pi(P(t))$ be the pixel which comes from the projection of $P(t)$ onto the retina Ω, where $\pi(\cdot)$ denotes the projection. Regardless of the knowledge of $\pi(\cdot)$, we can think of the invariance property which is induced by the trajectory $x(t)$. The brightness is approximatively constant on such a trajectory, that is, $b(x(t), t) = c$. Now, let $\varphi : \Gamma \to \mathbb{R}$ be any of the visual features of $\mathcal{O}$. The underlying assumption is that the computation of φ on any (x, t) is made possible mostly by the knowledge of $b(\cdot, t)$, that is, by the frame at time t. In a sense, $\varphi(x, t)$ is supposed to be a feature that is computable by a "nearly forward process" by involving a "quick" dynamical structure. A cognitive view of this assumption corresponds with the capability of describing "features" in a video nearly independently of time, on single frames, when focusing attention to a certain pixel. In any case, single frames do support visual information without needing to involve sequential processing. Apparently, this seems to be an unrealistic limitation but, on the opposite, it turns out to be one of the most important requirements of visual cognition. However, as it will become clear in the following, the need of developing tracking feature capabilities gives rise to an overall dynamical computational process that will be described in Chap. 5.

We can parallel the notion of optical flow for the brightness by introducing the corresponding notion for φ. Hence, the following consistency condition holds true:

$$\varphi(x_\varphi(t), t) = \varphi(x_\varphi(0), 0) = c_\varphi, \quad \forall t \in [0, T] \tag{3.5}$$

---

[17] The principle of material point consistency—MPI.

[18] Invariance takes place in portions of video delimited by saccadic movements.

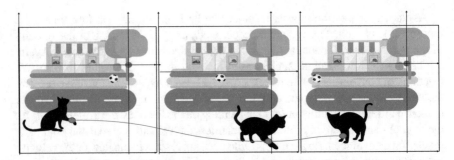

Fig. 3.2 Principle of material point invariance: When considering any pixel, the correspondent visual feature does not change during its motion, which will be considered either with respect to the computer retina or to the focus of attention

where $x_\varphi(t)$ denotes a trajectory of the associated feature φ and $c_\varphi \in \mathbb{R}$; if $x_\varphi(t) = x$ we let $v_\varphi(x, t) = \dot{x}_\varphi(t)$. See Fig. 3.2 for a sketch of the idea behind the consistency condition. It is important to notice that this principle relies on the underlying assumption of the joint existence of φ and v_φ on the corresponding trajectory x_φ that satisfies the condition stated by Eq. (3.5), which somewhat corresponds with thinking of φ as a signal which is propagating with velocity v_φ. In general, we can compute v_φ by a function which depends on $D\varphi(\cdot, t) = (\nabla\varphi(\cdot, t), \varphi_t(\cdot, t))'$. Like for the brightness, in general, the invariance condition generates an ill-posed problem. In particular, when the moving object has a uniform color, we noticed that brightness invariance holds for virtually an infinite number of trajectories. Likewise, any of the features φ is expected to be spatially smooth—nearly constant in small portions of the retina. This restores the ill-position of the classic problem of determining the optical flow that has been addressed in the previous section.[19] This time ill-position has a double face. Just like in the classic case of estimating the optical flow, v_φ is not uniquely defined. On top of that, now the corresponding feature φ is not uniquely defined, too. Basically, the MPI only states that material point invariance on the feature trajectories x_φ can be drawn jointly with the associated feature. Since the I Principle only establishes consistency on trajectories, we need to involve additional information for making the learning process well posed. A regularization process similar to the one invoked for the optical flow v—see Eq. (3.2)—can also be imposed for v_φ. However, unlike for the brightness, this time φ is not given. On the opposite, it is in fact characterizing the object, and it represents a "moving entity" of which we also want to measure the velocity. As discussed in the following, the functional structure of both φ and v_φ plays in fact a crucial role in the learning process and, correspondently, on its development.[20] The structure of φ is affecting the associated velocity v_φ and vice versa. In Sect. 4.4, we shall see that, as one begins thinking of the learning process behind v_φ, it is natural to pair such a process with the feature discovery, which leads

---

[19] Double face of ill-position in feature extraction.

[20] Neural-based structure for φ and v_φ.

to assume also a neural-based support for v_φ. The learning agent is supposed to jointly acquire both the features and their corresponding velocities.[21]

The formulation of the MPI raises the question on the precise meaning of the trajectory $x_\varphi(t)$. So far, we have considered the case in which an object is moving in front of a fixed retina. What if the object is fixed and the eyes and/or the body are moving? Moreover, it could also be the case that we are tracking a moving object while moving ourselves. This holds for machines as well. Since foveated animals move their eyes continually, their perception can naturally be understood by taking the FOA as a reference. As such, if we move, a stationary object turns out to be in motion along with the background. The bottom line is that *foveated animals are always in front of motion*, which conveys a continuous information flow. Of course, as pointed out later on, the motion invariance principle is independent of the frame of reference. The micro-saccadic movements enforce invariance on visually uniform patterns, while the object tracking provides a richer information on the object, since it is acquired in different areas of the background. Apparently, the body movement makes the problem considerably more involved, since the optical flow which arises does not distinguish the objects anymore. If we move uniformly on the right without moving our eyes, this clearly corresponds with a uniform optical flow on the retina with constant velocity on all the pixels in the opposite direction. It is interesting to see what happens if we propagate the feature consistency stated by MPI over the entire retina with respect to that constant velocity. Of course, in general, no feature meets such a criterion apart from the approximate satisfaction of the brightness stated by the invariance condition of Eq. (3.1). Basically, it looks like that if we focus on the certain pixel with velocity v, the corresponding uniform movement does not support any information for discovering visual features! However, as will be better seen in the following, the pairing of the features with their associated optical flow leads to establish an effective learning process also in this case, since the velocity fields of the features are not just reflecting the velocity associated with brightness invariance.

Equation (3.5) considers a single trajectory defined on the whole temporal horizon $[0, T]$, over which a very strict condition is imposed. In practice, we are interested in considering multiple trajectories, one of each point of the retina, that are locally defined in time. This leads to differential conditions that describe the local relationships over time of the way φ develops in the retina coordinates.[22] Formally, for any pair $(x, t) \in \Gamma$, let us overload the notation $x_\varphi(t)$ to represent the trajectory for which $x_\varphi(t) = x$ and so that $\dot{x}_\varphi(t) = v_\varphi(x, t)$. Given the optical flow v_φ, we say that φ is a *conjugate feature* with respect to v_φ provided that $\forall(x, t) \in \Gamma$; we have

$$\left(\varphi \bowtie v_\varphi\right)(x, t) := \frac{d\varphi(x_\varphi(t), t)}{dt} = \nabla\varphi(x, t) \cdot v_\varphi(x, t) + \varphi_t(x, t) = 0, \quad (3.6)$$

---

[21] What if the eyes and/or the body are moving? What if also the objects are moving? Take the FOA as a reference for movements!

[22] Conjugated features: We think of (φ, v_φ) as an indissoluble pair which must be jointly learned. We can detect the feature φ since "we can see" it moving at velocity v_φ.

where $\bowtie : \mathbb{R}^\Gamma \times \left(\mathbb{R}^\Gamma \right)^2 \to \mathbb{R}^\Gamma$ performs the mapping[23] $(\varphi, v_\varphi) \mapsto \varphi \bowtie v_\varphi$. Notice that if there is no optical flow in a given pixel $\bar{x}$, that is, if $v_\varphi(\bar{x}, t) = 0$ for all $t \in [0, T]$, then $\varphi_t(\bar{x}, t) = 0$. This means that the absence of the optical flow in $\bar{x}$ results into $\varphi(\bar{x}, t) = c_\varphi$ for all $t \in [0, T]$, which is the obvious consistency condition that one expects in this case. Likewise, a constant field $\varphi(x, t)$ in $C \subset \Gamma$ results into the conjunction $\varphi \bowtie v_\varphi = 0$ on C, independently of v_φ. Of course, since $b \bowtie v = 0$ we can legitimately state that b is in fact a conjugate feature—the simplest one that is associated with the brightness that is based on a single pixel. As for the optical flow associated with the brightness, also in this case, the regularization term (3.2), where v is replaced with v_φ, contributes to a well-posed formulation. Basically, (φ, v_φ) is an indissoluble pair that plays a fundamental role in the learning of the visual features that characterizes the object. Sometimes in what follows, we will drop the subscript φ of v_φ and x_φ when the notation is clear from the context.[24]

The I Principle has been stated by assuming the retina of the computer as the frame of reference. Foveated animals obviously experiment the MPI invariance by taking the fovea of their eyes as frame of reference. We can follow the same choice for machines. The trajectory $a(t)$ of the FOA makes it possible to set a frame of reference with axes parallel to those of the computer retina and center corresponding with the current position $a(t)$ of the FOA. In this new frame of reference, we always experiment motion, which suggests that there is always information coming from the invariance of the I Principle, including the case of still images and stationary visual agent.

Feature Grouping

As already noticed, when we consider color images, the brightness invariance can be considered for the separated components R, G, B. Interestingly, for a material point of a certain color, given by a mixture of the three components, we can establish the same brightness invariance principle, since those components can move with the same velocity. Said in other words, different features can share the same velocity. We can generalize this case by considering a group of features which share the same velocity. Hence, let $\varphi_i \bowtie v = 0 \ \forall i = 1, \ldots, m$ be, that is all features are conjugated with the same optical flow v, then we can promptly see that any feature φ of[25] $\text{span}(\varphi_1, \ldots, \varphi_m)$ is still conjugated with v. We can think of $\text{span}(\varphi_1, \ldots, \varphi_m)$ as a functional space conjugated with v. As already pointed out, the brightness is a quasi-invariant feature, but as we increase the degree of abstraction, we can aspire to increase the approximation of the conjugation of feature φ with its correspondent optical flow v_φ. Basically, large objects are perceived with a high degree of abstraction and are easier to track than their small parts. As already stated, the degree of ill-posedness of the determination of the classic optical flow is pretty high. A uniformly colored red rectangle, which is translating in front of us, makes it very difficult to

---

[23] Here with the symbol $\mathbb{R}^X$, we denote the set of all maps $f : X \to \mathbb{R}$.

[24] Frame of reference.

[25] Functional space $\text{span}(\varphi_1, \ldots, \varphi_m)$ represents a *feature group that is uniformly conjugated with the same optical flow*.

guess the velocity of the internal pixels of the rectangle. We cannot really track them! We can only make a guess that can be driven by the smoothness of the optical flow (3.2). However, we can easily follow the whole rectangle, just as the cat chases the mouse, that is regarded as a whole object. This way of conjugating features with the corresponding velocity seems to offer a formal support for the intuitive statement that we are tracking a moving object. A fundamental comment concerning the tracking of a feature vector $\phi = (\varphi_1, \ldots, \varphi_m)'$ which is very much related to the discussion on color tracking that is stated by Eq. (3.3). In this case, the invariance on ϕ yields[26]

$$\nabla\phi \cdot v + \phi_t = 0, \tag{3.7}$$

where $\nabla\phi \in \mathbb{R}^{m \times 2}$ is defined as $(\nabla\phi)_{ij} = \phi_{i,x^j} \equiv (\nabla\varphi_i)_j$. Notice that, if we consider the case in which the only scalar feature we are tracking is the brightness, then Eq. (3.7) boils down to a single equation with two unknowns (the velocity components). Differently, in the case of the feature group ϕ, we have m equations and still two unknowns. The dimension m of matrix $\nabla\phi$ can enforce the increment of its rank, which leads to a better positioning of the problem of estimating the optical flow v. Because of the two-dimensional structure of the retina, which leads to $v \in \mathbb{R}^2$ and since $\nabla\phi \in \mathbb{R}^{m \times 2}$, with $m \geq 2$ it turns out that any feature grouping regularizes the velocity discovery. In order to understand the effect of feature grouping, we can in fact simply notice that a random choice of the features yields rank $\nabla\phi = 2$. As a consequence, linear equation (3.7) admits a unique solution in v. Moreover, from Rouché–Capelli theorem we can also promptly see that in order to achieve the tracking of the feature group, the features need to satisfy

$$\text{rank } \nabla\phi = \text{rank}(\nabla\phi \mid -\phi_t).$$

However, this regularization effect of feature grouping does not prevent ill-positioning, since φ is far from being a random map. One the opposite, it is supposed to extract a uniform value in portions of the retina that are characterized by the same feature. Hence, rank $\nabla\phi = 1$ is still possible whenever the features of the group are somewhat dependent.[27] Feature groups that are characterized by their common velocity can give rise to more structure features belonging to the same group. This can promptly be understood when we go beyond linear spaces and consider for a set of indices $\mathcal{F}$

$$\begin{cases} \alpha = \displaystyle\sum_{j \in \mathcal{F}} w_j \varphi_j \\ \varphi = \sigma(\alpha). \end{cases} \tag{3.8}$$

Like for linear spaces, if $\varphi \bowtie v = 0$, then

---

[26] The regularization effect of *feature grouping*.
[27] Inheritance of conjugation.

$$\varphi \bowtie v = \sigma'(\alpha)\,(\nabla\alpha \cdot v + \partial_t\alpha) = \sigma'(\alpha)\sum_{j \in \mathcal{F}}(w_j \nabla\varphi_j \cdot v + w_j \partial_t\varphi_j).$$

From the conjugation equation $\varphi \bowtie v = 0$, we get

$$\varphi \bowtie v = \sigma'(\alpha)\sum_{j \in \mathcal{F}} w_j\left(\nabla\varphi_j \cdot v + \partial_t\varphi_j\right) = 0.$$

Hence, we conclude that if $\forall j \in \mathcal{F}$ we have $\varphi_j \bowtie v = 0$ then also the feature φ defined by Eq. 3.8 is conjugated with v, that is, $\varphi \bowtie v = 0$. We can promptly see that vice versa does not hold true. Basically, the inheritance of conjugation with v holds in the direction toward more abstract features. Of course, the feedforward-line recursive application of the derivation stated by Eq. (3.8) yields a feature φ that is still conjugated with v. In this case, we use the notation $\vdash \varphi$ to indicate such a data flow computation.

Computational Models for (φ, v_φ)

The notion of conjugate features does require the involvement of the specific computational structure of φ. In general, we assume that φ and v_φ turn out to be expressed by

$$\begin{aligned}
\frac{\partial\varphi}{\partial t}(x, t) + c\varphi(x, t) &= c\alpha_\varphi\big(b(\cdot, t), a(t), t\big)(x); \\
\frac{\partial v_\varphi}{\partial t}(x, t) + cv_\varphi(x, t) &= c\alpha_v\big(D\varphi(\cdot, t), a(t), t\big)(x),
\end{aligned} \tag{3.9}$$

where $\alpha_\varphi \colon \mathbb{R}^\Omega \times \Omega \times [0, T] \to \mathbb{R}^\Omega$, $\alpha_v \colon \mathbb{R}^\Omega \times \Omega \times [0, T] \to (\mathbb{R}^\Omega)^2$, $t \mapsto a(t)$ is the trajectory of the focus of attention and $c > 0$. Here, c is supposed to drive very quick dynamics, so as if the input $b(\cdot, t)$ is kept constant we end up quickly into the associated stationary condition $\varphi(x, t) = \alpha_\varphi(b(\cdot, t), t)(x)$. Basically, this corresponds with performing a "nearly forward computation." In human vision, the system dynamics is expected to be compatible with classic frame per second frequency. The same computational structure is assumed for the conjugated velocity $v_\varphi(x, t)$. The explicit dependence on t in α_φ and α_v is due to the underlying assumptions that those functions undergo a learning process during the agent's life. In Eq. (3.9), it is assumed that the computation of the features at a point $x_0 \in \Omega$ requires the knowledge of the brightness on the whole retina. However, this assumption can be relaxed by requiring that we pass to the function α_φ the restriction $b(\cdot, t)|_{\Omega_\varphi(x_0)}$ of the brightness to a subset $\Omega_\varphi(x_0) \subset \Omega$. In doing so, the computation of $\varphi(x_0, t)$ makes it possible to take decisions on the actual portion of the retina $\Omega_\varphi(x_0)$ that can be virtually considered when computing the feature on pixel x_0. This is somewhat related to the convolutional neural networks' computational scheme, and it will be specifically described in Chap. 4.

The computational structure of v_φ is suggested by its conjugation with φ, where we see the dependence on $D\varphi$. This is related to the classic idea early illustrated by Lukas and Kanade in [68] for the optical flow, where its determination is based on stating the conjugation condition on a receptive field $\Omega_b(x)$ centered on the pixel x where we focus attention.

Of course, the conjugation of φ and v_φ does involve the choice of both α_φ and α_v. From Eqs. (3.6) and (3.9), we get the following consistency condition on φ_t

$$c^{-1}\nabla\varphi(x,t) \cdot v_\varphi(x,t) = \varphi(x,t) - \alpha_\varphi\big(b(\cdot,t),a(t),t\big)(x). \qquad (3.10)$$

As we can see, for large values of c, this corresponds to a quasi-forward computation.

Regularization for Determining (φ, v_φ)

We have already discussed the ill-posed definition of features conjugated with their corresponding optical flow. Interestingly, we have shown that a feature group $\phi = (\varphi_i, \dots, \varphi_m)'$, which shares the velocity inside the group, exhibits an inherent regularization that, however, does not prevent ill-positioning, especially when one is interested in developing abstract features that are likely constant over large regions in the retina. Let us assume that we are given n feature groups $\phi^i, i = 1, \dots n$, each one composed of m_i single features (m_i-dimensional feature vector) $\phi^i \in \mathbb{R}^{m_i}$. Furthermore, let us assume that each group ϕ^i shares the same velocity v^i with all the features of the group. Let us also denote with $\varphi = (\phi^1, \dots, \phi^n)$ and with $\mathbf{v} = (v^1, \dots, v^n)$.[28] We can impose the generalization of the smoothness term used for the classic optical flow to the velocities of the features by keeping the following term small:

$$E = \frac{1}{2}\sum_{i=1}^{n} \int_\Gamma \left(\|\nabla v^i(x,t)\|^2 + \lambda_\varphi|\phi^i(x,t)|^2 + \lambda_\nabla\|\nabla\phi^i(x,t)\|^2\right) d\mu(x,t). \qquad (3.11)$$

Here, $\|A\|^2 = \sum_{i,j} A_{ij}^2$ and $\lambda_\varphi, \lambda_\nabla$ are positive constants that express the relative weight of the regularization terms and the measure μ is an appropriately weighted Lebesgue measure on $\mathbb{R}_x^2 \times \mathbb{R}_t$: $d\mu(x,t) = \exp(-\theta t)g(x - a(t)) dx dt$. Here, we have $\theta > 0$, so as $\exp(-\theta t)$ produces a decay as time goes by. Function g is nonnegative, and it has a radial structure and decreases as we get far away from the origin. It takes on the maximum for $x = a(t)$, where $t \mapsto a(t)$ denotes the coordinates of the focus of attention. Notice that E is a functional of the pair (φ, v); that is, once it is given, we can compute $E(\varphi, \mathbf{v})$. When comparing this index with the one used to regularize the classic optical flow (see Eq. (3.2)), we can see a remarkable difference: E is also a function of time! Why is that necessary? While the brightness is given, the features are learned as time goes by, which is just another facet of the feature–velocity conjugation. It is worth mentioning that E does only involve spatial smoothness whereas it does not contain any time regularization term. As it will become more clear in Chap. 5, such a regularization is in fact introduced into the weight of the neural networks that are used to implement Eq. (3.9). We can also introduce a form of temporal regularization by upper-bounding $|v^i|$, since it directly affects the conjugated features, thus limiting $\partial_t\phi^i$. There is also another difference with respect to the classic optical flow (see Eq. (3.2)): There is also a penalizing

---

[28] Default "null" and spatial smoothness.

term $(1/2)|\phi^i|^2$ which favors the development of $\phi^i = 0$. Of course, there is no such requirement in classic optical flow, since b is given. On the opposite, the discovery of visual features is expected to be driven by motion information, but their "default value" is expected to be null. We can promptly see that the introduction of the regularization term (3.11) does not suffice. The conjugation ⋈ is in fact satisfied also by the trivial constant solution $\varphi = c_\varphi$.[29]

Important additional information comes from the need of exhibiting the human visual skill of reconstructing pictures from our symbolic representation. At a certain level of abstraction, the features that are gained by motion invariance possess a certain degree of semantics that is needed to interpret the scene. However, visual agents are also expected to deal with actions and react accordingly. As such, a uniform cognitive task that visual agents are expected to carry out is that predicting what will happen next, which is translated into the capability of guessing the next incoming few frames in the scene. We can think of a predictive computational scheme based on the ϕ^i codes which works as follows

$$\frac{\partial y}{\partial t}(x, t) + cy(x, t) = c\alpha_y\big(\varphi(\cdot, t), t\big)(x). \tag{3.12}$$

where $\alpha_y : \big(\mathbb{R}^\Omega\big)^n \times [0, T] \to \mathbb{R}^\Omega$. In particular, the prediction y needs to satisfy the condition established by the index[30]

$$R = \frac{1}{2} \int_\Gamma \big(y(x, t) - b_t(x, t)\big)^2 \, d\mu(x, t). \tag{3.13}$$

Of course, as the visual agent gains the capability of predicting what will come next, it means that the developed internal representation based on features φ cannot correspond with the mentioned trivial solution. Interestingly, it looks like visual perception does not come alone! The typical paired skill of visual prediction that animals exhibit helps regularizing the problem of developing invariant features.[31] More sophisticated predictions can rely on a dynamical system to predict the features and the velocities. However, this book makes the fundamental assumption of disregarding dynamical processes for the representation of $(\varphi, \mathbf{v})$. The computational models (3.9) and (3.12) exhibit in fact a "nearly forward structure." This assumption is somehow related to the hypothesis of picture perceivable actions that is introduced in detail in Sect. 3.6 concerning the notion of affordance. It is claimed that system dynamics is needed for high-level human-like cognitive processes (e.g., understanding actions like to tie the perfect bow or solving Rubik's cube). On the opposite, most animal and human actions do not need to involve system dynamics but simply the dynamics connected with the optical flow of the features. In any case, the prediction carried out by y is assumed to be rich enough, a process that vaguely reminds us of classic predictive

---

[29] What happens in the next frames?

[30] The prediction tasks consist of discovering an output y such that: $y(\cdot, t) \simeq b_t(\cdot, t)$.

[31] No system dynamics apart from that inherited by motion!

coding [77]. A recent survey on deep learning techniques on the construction of the prediction scheme stated by Eq. (3.13) has been published in [94].

A few comments are in order concerning the regularization term and the functional which controls prediction.

- *Features φ are expected to express visual properties with different degrees of abstractness that is useful for perception as well as for action, since we need to gain prediction skills.*

  We must bear in mind that those features are somehow characterizing symbolic properties of a given object. In particular, they are expected to embrace a non-negligible portion $\Omega_\varphi(x)$ of the retina from which one is expected to be able to provide descriptions. As we collect all those "descriptions," we expect that they are sufficient to reconstruct the retina. Of course, since they are expressed by real numbers, they address hidden semantics that we could find it hard to describe. Yet, for any pixel, we collect formal descriptions that drive prediction. For example, in order to help intuition, one can think of a red cat with bald spots and wet fur. Notice that it is a truly global description of a whole object. It could also be the case that the cat is given a more articulated symbolic description which refers to his head or to any other part of the body. This kind of descriptions might be enough to identify a cat in a limited population, and consequently, it could be enough for the generation of the corresponding picture. The φ codes are basically of global nature, and therefore, the predictive reconstruction cannot exhibit trivial solutions. Clearly, the $\varphi = c_\varphi$ which satisfies motion invariance is not acceptable since it does not reconstruct the input. This motivates the involvement of prediction skills typical of action that, again, seems to be interwound with perception.

- *E and R are functional, where the dependency on $\Omega_\varphi(x)$ is the sign of a corresponding spatiotemporal dependency induced by the FOA.*

  As pointed out in the list of motivations for choosing foveated eyes in item (2) of Sect. 2.3, there are good reasons for choosing variable resolution when considering the temporal structure of the computation and the focus of attention mechanism. Moreover, still in the same list, the claim in (6) puts forward the limitations of weight sharing with the need of disambiguating decisions in different positions in the retina. Concrete computational mechanisms for carrying out this kind of computation will be described in Sect. 4.4 by discussing deep neural networks.

- *Unlike what we have seen for Horn and Schunck variational formulation of the optical flow, here the weights depend on the environmental time.*

  This is in fact at the basis of the variational structure of the learning processes considered in this book.

Even though the learning process will be described in Chap. 5, here we want to provide a preliminary formulation of learning, which has its roots in related recent studies [12, 15]. Let S be the following functional[32]

---

[32] Formulation of learning the pair (φ, v) according to the principle of least cognitive action.

$$S(\varphi, \mathbf{v}) = \lambda_E E(\varphi, \mathbf{v}) + \lambda_R R(\varphi, \mathbf{v}) + \frac{1}{2} \sum_{i=1}^{n} \int_{\Gamma} |\phi^i \bowtie v^i(x, t)|^2 \, d\mu(x, t), \quad (3.14)$$

that, throughout this book, is referred to as the *cognitive action*. Here, λ_R and $\lambda_E > 0$ are the regularization parameters. Learning to see means to discover the indissoluble pair $(\varphi^\star, \mathbf{v}^\star)$ such that

$$(\varphi^\star, \mathbf{v}^\star) = \arg\min_{(\varphi, \mathbf{v})} S(\varphi, \mathbf{v}). \quad (3.15)$$

Basically, the minimization is expected to return the pair $(\varphi, \mathbf{v})$, whose terms should nominally be conjugated. The case in which we reduce to consider only the brightness, that is, when the only φ is b, corresponds with the classic problem of optical flow estimation. Of course, in this case the term R is absent and the problem has a classic solution. Another special case is when there is no motion, so as the integrand $(\phi^i \bowtie v^i)^2$ is simply null $\forall i = 1, \ldots, n$. In this case, the learning problem reduces to the unsupervised extraction of features φ. In classic pattern recognition, the world is dramatically characterized by the underlying principle that perceptual vision is in fact a classification process. One defines the classes in advance and the subsequent learning process is affected by the choice of which classes are in fact characterizing the given visual world. Hence, the introduced classification task is somewhat biasing the interpretation of the given visual environment. When the subtleties of language become an important issue to consider in visual descriptions, it becomes early clear that one cannot reasonably define the classes in advance. Ordinary visual descriptions consist of linguistic statements, which are well beyond object categorization, with classes defined in advance. The compositionality of language suggests that the visual learning processes centered around classification are likely specializing toward a sort of artificial process. The fundamental requirement of prediction stated by (3.13) leads to reconstruct video, regardless of any prior categorization that has been eventually assumed on the objects.[33]

The well-posedness of learning according to (3.15) for the pair $(\varphi, \mathbf{v})$ can be established also without involving the I Principle by imposing motion invariance. The prediction and the regularization terms yield in fact an appropriate definition of the minimum. The codes φ and the associated velocities v_φ are extracted with the purpose of predicting b_t and to gain uniform values as much as possible for both variables. The introduction of motion invariance, however, produces the fundamental effect of making the decisions consistent with motion. It turns out that in the given visual environment, we need not to express the solution over the real field, but on the space of countable features. It is in fact the integer which identifies the feature that can be used for representing the information that has been learned. In a sense, $(\varphi, \mathbf{v})$ are countable symbols over the given visual environment. While this can always be assumed in any neural network, we need to involve visual environments without the bias of classification for conquering visual interpretation. Again, the fundamental difference

---

[33] The universal property of codes ($\varphi^\star$ gained along with $\mathbf{v}^\star$).

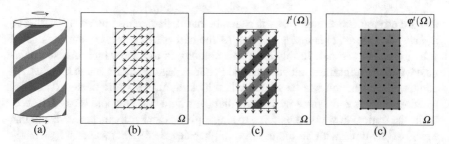

Fig. 3.3 Barber's pole example. **a** The 3-D object spinning counterclockwise. **b** The 2-D projection of the pole and the projected velocity on the retina Ω. **c** The brightness of the image and its optical flow pointing upward. **d** A feature map that responds to the object and its conjugate (zero) optical flow

with respect to typical deep learning approaches to computer vision is that features are motion consistent, which in turn minimizes the amount of information needed to perform visual prediction.[34] The satisfaction of $\varphi \bowtie v_\varphi = 0$ corresponds with $d\varphi/dt = 0$ along moving trajectories. Notice that this condition does not prescribe the constant value of φ uniformly on the retina, but only on trajectories. When $v^i = 0$, the conjugation yields constant in time ϕ^i (keep the decision on the pixel). While we expect that the $\varphi \neq 0$ be "strongly met" when the feature is active, we expect $\varphi \simeq 0$ when it is non-active. Such a "contrastive condition" is not the outcome of invariance, since invariance only enforces $\varphi = c$. However, the additional regularization term penalizes $\varphi \neq 0$, which is the value that the feature takes on when it is non-active. Clearly, non-null values of the features are sustained by need of performing frame prediction. We end up into the conclusion that any point of the retina, for any frame, either belongs to the partition where the feature is active or non-active, with constant value for φ with both cases. Motion invariance suggests that the storage of the representation of φ collapses to a finite representation. Notice that all this analysis holds for a functional space which admits abrupt transition from active to non-active values. However, the approximation capabilities of neural nets make it possible to get very close to this ideal condition.

Barber's Pole and Degree of Invariance

Now, let us consider another example of optical illusion produced by objects themselves. This is very instructive in order to get an intuition on the meaning of velocity of features. The following classic *barber's pole* offers an intriguing example to understand the importance of introducing appropriate optical flows for the visual features. As the pole rotates counterclockwise, it is hard to visually track the single pixels. Clearly, since they are the projections of rotating material points, the velocity v of a single pixel is a horizontal vector that can easily be computed when the rotational speed of the pole is given, but its visual perception does not arise, since the resulting optical flow is in fact an optical illusion. Now, as we focus attention on the stripes of the pole, surprisingly enough, we can easily track them; they have an apparent

---

[34] Motion invariance and MDL.

upward motion. Now, suppose φ_r is a feature that characterizes the red stripe, that is, $\varphi_r(x, t) = 1$ iff (x, t) is inside a stripe. As the barber's pole rotates, the conjugated velocity v_{φ_r} is vertical. In this case, this velocity is the one which is also conjugated with the brightness under Horn and Schunck regularization stated by (3.2). An additional abstraction can be gained when looking at the whole object: Its conjugated velocity is zero, since we assume that the whole barber's pole is fixed. Hence, while the optical flow corresponding to single pixels can hardly be perceived, in this case, two features with an increasingly higher degree of abstraction are perfectly perceivable with their own conjugated velocities. This is very much related to the mentioned chasing skill of the cat! In real-world scenes, the nice conjugation of the vertical velocity with the stripes in the barber's pole can hardly be achieved. Unlike the case of the brightness, however, we can always think that a feature does represent a more distinguishing property that is associated with the object. The postulation of the existence of those features corresponds with stating their invariance. In general, one can believe in the indissoluble conjunction of features and velocities, stated by $\varphi \bowtie v_\varphi = 0$, and, consequently, on their joint discovery. As we see the barber's pole stripes, we also track their vertical movement when the pole is rotating. The perception of the vertical movement comes with the perception of the corresponding stripe and vice versa.

3.6 The Principle of Coupled Motion Invariance

As pointed out in Sect. 3.3, visual perception encompasses of both object identification and affordance. The MPI (I Principle) only involves identification. In this section, we begin addressing seminal Gibson's contributions [37, 38], which have inspired scientists for decades. His studies in the field of psychology have been contaminating the community of computer vision (see, e.g., [34]) also in the era of deep learning [45]. Object affordance is interwound with correspondent actions, but studies on these topics are often carried out independently one of each other in computer vision. Interestingly, the interpretation of actions is typically regarded as a sequential process. By and large, actions are perceived as the outcome of a sequence of frames. Studies in deep learning for action recognition somehow aggregate features and optical flow. Sequences are taken into account mostly by combining convolutional neural networks and long short-term memory recurrent neural networks [93], which are in charge for sequential processing.[35]

Interestingly enough, humans can promptly perceive many common actions from single pictures, which has clearly an impact in the notion of affordance! For example, we can easily recognize the act of jumping, as well as the act of pouring a liquid into a container from a single picture, which can also be enough for understanding that somebody is eating or drinking. The act of taking a penalty kick or scoring a basket is also *picture perceivable actions*. Hence, it looks like there are actions that

---

[35] Picture perceivable actions and the transmitted affordance.

are perceived from single pictures without presenting video clips. It is important to realize that when shifting attention to actions, we need to focus on their meaning that is paired with sophisticated human experiences. While we can recognize from a single picture the act of pouring a liquid into a container, which nicely yields the correspondent affordance, we cannot recognize the act of solving Rubik's cube (while we can say that the cube is being manipulated by someone). A single picture can convey suspiciousness on the acts of killing, but it cannot suffice to drive the conclusion on the murder. Similarly, a photograph in which I am trying to make myself the bow does not really correspond with the action labeled as "how to tie the perfect bow," since the photograph could just trace my beginning act which might not result in a successful tie of the perfect bow! The acts of solving Rubik's cube, killing, and tying the bow do require video clips for their interpretation, whereas other actions, like running, jumping, grazing the grass, and pouring a liquid into a container, are picture perceivable. The natural flow of time conveys a sort of direction to actions like seating and standing up; they are similar, yet somewhat opposite in their meaning. The same holds for the act of killing and cutting the bread. We can hit with the knife or remove it from an injured person. Clearly, the simple act of removal can support an important difference: The person who is acting might be the killer, who is just removing the knife for hitting again, or might be a policeman who has just been notified of the murder. Those actions definitely need a sequence-based analysis for their interpretation, and they also support a different affordance, which has in fact a truly global nature. On the opposite, whenever actions are perceivable from single pictures the corresponding affordance has a local nature. Regardless of this difference, in general, more actions can refer to the same object (e.g., the chair and the knife), which means that the process of object recognition generally involves multiple affordances. The notion of chair clearly arises from both the acts of seating and standing up, but it also emerges when we use it for reaching an object.[36] The same holds true for many other objects. This discussion suggests that we transfer affordance to objects even with picture perceivable actions. Moreover, while some actions, like cutting the bread and seating on a chair, could still be confused when interpreted by a single picture, it can be enough to convey affordance to the bread and the chair, respectively. Single pictures of humans carrying out actions on an object do convey affordance. While we can argue on the degree of semantic information which is transmitted, in general, those pictures are definitely important for understanding the function of the object. While solving Rubik's cube does require the frames of the entire video where the act is shown, also in this complex case a single picture on cube manipulation does convey precious affordance. The cube can be recognized as an object that can be manipulated, regardless of the fact that the manipulation actually leads to the solution. When considering different actions on a certain object, we can immediately realize that they all convey affordance. For example, pictures of the acts of cutting or breaking bread, both transfer an affordance that is useful for recognizing bread. This discussion strongly pushes the importance of the *local notion of affordance* and suggests that more abstract interpretations

---

[36] Affordance is conveyed also when the action is not perceivable by a single picture.

corresponding to portions of video may not play the central role in the transmission of affordance. They are in fact conveying a more abstract and high-level information that can be interpreted only by sophisticated reasoning models. This goes beyond the basic principle of perceptual vision and action control, and it is beyond the scope of this book.

Second Principle of Perceptual Vision: Coupled Motion Invariance (CMI)

Like for object identification, the notion of affordance becomes more clear when shifting attention at pixel level, which is in fact what foveated animals do. First, we need to assume a sort of animal-centric view since object affordance is better understood when it is driven by an animal.[37] This assumption is especially useful to favor the intuition behind the II Principle of perceptual vision, but it will be removed and the analysis will be extended to the case of visual features, thus paralleling the statement of the MPI principle. We shall begin discussing interactions between animals and objects that can be identified by corresponding codes but, as stated in the following section, the ideas herein presented can be given a more general foundations in terms of the interaction between objects in visual environments. Without limitation of generality, suppose we are dealing with human actions, and let us consider a pixel x where both a person, with identity $p(x, t)$, and an object, with identity $o(x, t)$, are detected. As already noticed, in general, we can also assume that the object defined by $p(x, t)$ is not necessarily referring to a person, but to tools typically used in human actions. For instance, $p(x, t)$ could be the identifier of a hammer which is used to hang a picture. Interestingly, a pixel could contain the identity of the hammer, the picture, and the fingers of a person. In general, it could be the case that more than one object is defined at (x, t), which depends on the extent to which we consider the contextual information. In that case, the action also gives rise to a corresponding number of optical flows at (x, t) that are conjugated with the coupled objects. For the sake of simplicity, suppose the optical flow defined by $v_p(x, t)$ in pixel x at time t comes from the person who is providing the object affordance.[38] Following the conditions under which the I Principle of visual perception has been established, let us consider a single movement burst delimited by saccadic movements. For any $(x, t) \in \Gamma$, we can introduce the *coupling relation* $\between$ between objects as follows:

$$p \between o(x, t) := \gamma(p(x, t)) \wedge \gamma(o(x, t)).$$

Here, γ returns the Boolean decision behind o and p; it is typically $\gamma : [0, 1] \rightarrow \{0, 1\}$. A value is returned that is based on thresholding criteria. Of course, depending on the measure of $\mathscr{C}_{p \between o} = \{(x, t) \in \Gamma : p \between o(x, t) = 1\}$ we can experiment a different degree of object coupling, so as one can reasonably establish whether the field interaction between $p(x, t)$ and $o(x, t)$ is significant. Given $\epsilon > 0$, we say that

---

[37] As it will be better seen in the following, a more general view on object affordance suggests that it can come also from another object which is moving, but the following analysis clearly holds regardless of who is transferring the affordance.

[38] II Principle of Perceptual Vision: Coupled Motion Invariance (CMI): the person identifier $p(x, t)$, paired with object identifier $o(x, t)$.

the coupling $o \, \langle \, p$ is ϵ-significant provided that[39] $\mathcal{L}^3(\mathscr{C}_{p\langle o}) > \epsilon$ and write $o \, \langle_\epsilon \, p$.[40] Whenever this happens, it turns out to be convenient to introduce the *degree of affordance* $\alpha_{op} \in \mathbb{R}$ of o conveyed by p. Unlike the information on the object identification, the degree of affordance transmits the information on how the person defined by p is using o. We are now ready to establish the coupling motion invariance principle as follows[41]:

$$\text{CMI}: \qquad \alpha_{op} \bowtie (v_o - v_p) = 0. \qquad\qquad (3.16)$$

This statement is very related to the I Principle. If the interacting object p is not moving, that is, $v_p = 0$, then CMI reduces to $\alpha_{op} \bowtie v_o = 0$, which is reminiscent of the constraint $o \bowtie v_o = 0$. Hence, in this case α_{op} resembles the identifying code o. Now, there are good reasons for thinking of the two principles separately and keep α_{op} and p separated. The I Principle is about object identification, whereas the II Principle is about affordance, and they are in fact conveying a different, yet complementary, information! A visual agent needs to recognize chairs, but he also needs to identify the specific type of chair, and he must also be capable of inducing that the specific object is in fact part of a more general category. Such a category may reflect a few distinctive visual properties of an object which also support its function. This is in fact the reason for keeping o and α_{op} separated. The difference emerges when p interact with o also by its velocity v_p. In this case, Eq. (3.16) still states the motion invariance property of o, with respect to the relative velocity $v_o - v_p$. It is worth mentioning that this must be interpreted in the local sense and that CMI and MPI are essentially stating the same property of motion invariance of o, but the fundamental difference is the use of the relative velocity. The presence of p with velocity v_p is in fact the indication of the affordance conveyed by p to o, which results in the different references where the object velocity must be measured. When the fixed reference is chosen, we are electing the distinctive feature o of the object, whereas the presence of interacting objects leads to express invariance with respect to their reference.

It is worth mentioning that the same person p can propagate the affordance to objects with a different identity o_1, o_2. For example, we manipulate Rubik's cube as well as a hammer, or a knife. This induces the relation $o_1 \overset{p}{\sim} o_2$ which holds if

$$(p \, \langle_\epsilon \, o_1) \quad \text{and} \quad (p \, \langle_\epsilon \, o_2). \qquad\qquad (3.17)$$

We notice in passing that this relationship does not assume a specific semantics in the action carried out by p. While in the previous example we were considering the act of manipulation on different objects, the above stated field interaction might also hold for touching, pushing, or shaking. Of course, this analysis could be extended

---

[39] $\mathcal{L}^3$ here is the Lebesgue measure on $\mathbb{R}^3$.

[40] *Degree of affordance* $\alpha_{op} \in \mathbb{R}$ of o conveyed by p.

[41] Coupled motion invariance.

if one can classify actions, but this is not considered in this book.[42] Two persons, defined by p_1, p_2, can also transfer the same affordance to o, which is represented by the dual relation with respect to (3.17) $p_1 \overset{o}{\sim} p_2$ which means:

$$(p_1 \, \langle_\epsilon \, o) \quad \text{and} \quad (p_2 \, \langle_\epsilon \, o). \tag{3.18}$$

The abstraction of Eq (3.17) leads to the notion of action. For example, the same person can manipulate different objects. This characterizes a certain action of p which holds for a certain class of objects. We can in fact manipulate Rubik's cube as well as the knife and many other objects. The abstraction of Eq. (3.18) consists of considering the case in which object affordance is conveyed independently of the person. In this case, Eq (3.17) can be replaced with $o_1 \sim o_2$. Overall, when a visual learning environment along with multiple interactions is considered we can create a partition on object identifiers O which yields the quotient space $Q = \mathcal{O}/\sim$ where the equivalent classes can be thought as the class of the object.[43]

Now, let us assume that in a visual environment there is collection $\mathcal{P}_o$ of person identifiers who are coupled with o. The joint coupling with all people in $\mathcal{P}_o$ leads to think of the corresponding affordance $\alpha_{o\mathcal{P}}$. However, as $\mathcal{P}_o$ becomes large enough it does characterize the "living environment" and it makes sense to think of the *inherent affordance notion* α_o for which the following condition holds true:

$$\alpha_o \bowtie (v_o - v_j) = 0 \quad \forall j \in \mathcal{P}_o. \tag{3.19}$$

For example, when considering the notion of knife, we can think of its inherent affordance which is gained by the act of its manipulation by a virtually unbounded number of people. Of course, Eq. (3.19) can only be given a formal meaning, since in real-world environments the notion of inherent affordance can only be approximatively gained. Unlike the single pairing of o with p where the affordance is modeled by α_{op}, here the inherent affordance α_o involves a collection of constraints that is logically disjunctive over $\mathcal{P}_o$.[44] This can be expressed by the corresponding risk

$$A = \sum_{j \in \mathcal{P}_o} \left(\alpha_o \bowtie (v_o - v_j) \right)^2. \tag{3.20}$$

The way we conquer a formal notion of affordance, stated by Eq. (3.16), works for both *attached* and *detached objects*, since their movement is not necessary. The affordance comes from people who are using the object. However, the way people interact with attached and detached objects is definitely very different. We experiment an interaction with objects that can be directly grasped that is far away from the interaction with big fixed objects. While hammers can be grasped and moved so

---

[42] The basic ideas are the same whenever in the coupling $o \, \langle \, p$, the object identifier p is replaced with an action field.

[43] Inherent affordance.

[44] Dealing with attached objects.

as their motion with respect to the environment can be used for conquering their identity, a building can neither be grasped nor moved. Yet, humans and attached objects can still interact so as they can get the affordance from human movements. As we look out the window or open the shutters, the corresponding movement transfers an affordance. Something similar happens as we walk close home. Even though a chair is clearly detached, it is typically fixed as we sit down or stand up. Of course, it can itself slightly move, but the affordance is basically gained from the person who is using it.[45]

Because of the structure of the optical flow generated during object–human interaction, it is quite obvious that the identity of detached objects can better be gained. On the opposite, for both attached and detached objects the affordance can be similarly gained. There is still a difference, since in detached objects the optical flow comes from both humans and objects. Hence, it looks like attached objects are likely more difficult to learn, since the development of correspondent identification features is more difficult than for detached objects.

3.7   Coupling of Vision Fields

As discussed in the previous two sections, the I and the II Principles come from the need of developing two different, yet complementary, visual skills: object identification and affordance. A certain discrepancy in expressing the two principles may not have gone unnoticed. While the I Principle has been expressed on object features, the II Principle has involved object identifiers, that is, higher-level concepts. Obviously, this discrepancy could be reconciled by also stating the I Principle directly in terms of object identifiers, so as in both cases we directly involve objects as atomic entities. That would be fine, but the simple statement of the I Principle on the object feature suggests that both principles might be interpreted in terms of a *vision field theory*, the purpose of which is that of studying the interactions between features and velocities. As already noticed, the II Principle does not really need to make explicit the semantic of the actions that transfer affordance. There is one more reason to focus on features instead of objects, which is connected with the chicken–egg dilemma discussed in Sect. 1.5: While we can define the velocity of features by involving video information only, as it will be pointed out in Sect. 6.1, this is not sufficient to fully capture the meaning of attached objects.[46]

Hence, a question arises on whether or not affordance can be gained also by ignoring the object identity which transfers affordance. Notice that while it is gained by motion invariance of the same object, the essence of affordance also comes from a motion invariance condition that involves the motion of who is using the object. However, in the statement of the II Principle there could be hidden semantics in the

---

[45] Attached and detached objects involve a different optical flow; identification is favored for detached objects.

[46] Affordance is also conveyed by visual features with hidden semantics.

entities which convey affordance. As such, we can think of visual features and of a more general view that involves their coupling.[47] Let us consider the collection of features $\{ (\phi^i, v^i) : i = 1, \ldots, n \}$ which arises from a certain visual environment. In what follows, we regard ϕ^i as scalar features (i.e., in our previously introduced notation $m_i \equiv 1$ for $i = 1, \ldots, n$). In general, these features cannot be easily associated with object identifiers, but we can consider visual features as interacting fields. While the conjugation of self-coupling $\phi^i \bowtie v^i = 0$ of the I Principle favors the feature identification, the coupling $\phi^i \lozenge \phi^j$ with $i \neq j$ can come from the interaction of either different objects or within the same object. The first case corresponds with what we have already discussed concerning object affordance. In the second case, we are still in front of a true field interaction between features; for example, we can move our hands while walking, which likely establishes a vision field where the features corresponding to the head are somewhat coupled with the velocities associated with the hands. Interestingly, while making these examples helps the intuition and it is definitely stimulating, the mathematical structure behind vision field coupling speaks for itself. Hence, we replace the given definition of affordance α_{op} by introducing the field $\psi_{ij} : \Gamma \rightarrow \mathbb{R}$, which indicates the affordance that feature ϕ^j transfers to feature ϕ^i.[48]

Based on this premise on feature coupling, we can provide a more general statement of the II Principle. When considering this general perspective, instead of assuming that humans are transferring affordance to a fixed object, one must take into account a velocity field for each interacting feature. Under these conditions, the II Principle can be stated as:

$$\mathrm{CMI - bis}: \qquad \psi_{ij} \bowtie (v^i - v^j) = 0 \qquad 1 \leq i, j \leq n. \qquad (3.21)$$

Like in the statement (3.16) of the II Principle which refers to objects, ψ_{ij} is conjugated with velocity $v^i - v^j$, which is in fact the relative velocity of feature ϕ^i in the reference of feature ϕ^j, which transfers the affordance. Notice that if affordance feature ψ_{ij} does not receive any motion information from ϕ^j, since $v^j = 0$, then the consequent condition $\psi_{ij} \bowtie v^i = 0$ is in fact a self-motion invariance corresponding with the I Principle. However, it can better be interpreted as a *regularization statement* for ψ_{ij} that leads to express consistency on the affordance interpretation of the feature itself. In other words, it can be interpreted as a feature self-coupling, that is, as a reflective property of ψ_{ij}.[49] As already stated in the previous section, when the visual environment is given, as time goes by, the object interactions begin obeying statistical regularities and the interactions of feature ϕ^i with the other features become very well defined. Hence, the notion of ψ_{ij} can be evolved toward the *inherent affordance* ψ_i of feature φ_i. Based on Eq. (3.19), we define the inherent feature affordance as the function ψ_i which satisfies

---

[47] ψ_{ij} is the *feature affordance* that ϕ^j transfers to ϕ^i: It comes from the coupling $\phi^i \lozenge \phi^j$.

[48] Vision field-based restatement of the II Principle and symmetry of ψ_{ij}.

[49] Inherent affordance ψ_i: It comes from the interactions with all the environmental features.

$$\psi_i \bowtie (v^i - v^j) = 0, \qquad \forall j = 1, \ldots, n, \qquad 1 \leq i \leq n. \tag{3.22}$$

This reinforces the meaning of inherent affordance, which is in fact a property associated with ϕ^i while living in a certain visual environment. As a consequence, the identification feature ϕ^i pairs with the corresponding affordance feature ψ_i, so as the enriched vision field turns out to be defined by $\mathcal{V} = (\phi, \psi, v)$. In a sense, ψ can be thought of as the abstraction of ϕ, as it arises from its environmental interactions. A few comments are in order concerning the visual field.

- The pairing of ϕ^i and ψ_i relies on the same optical flow which comes from ϕ^i. This makes sense, since the inherent affordance is a feature that is expected to gain abstraction coming from the interactions with other features, whereas the actual optical flow can only come from identifiable entities that are naturally defined by ϕ^i.
- The inherent affordance features carry significantly redundant information. This can be understood when considering especially high-level features that closely resemble objects. While we may have many different chairs in a certain environment, one would expect to have only a single concept of chair, which in fact comes out as the affordance of an object used for seating. On the opposite, ψ assigns many different affordance variables that are somewhat stimulated by a specific identifiable feature. This corresponds to thinking of these affordance features as entities that are generated by a corresponding identity feature.
- The vision field $\mathcal{V}$ is the support for high-level decisions. Of course, the recognition of specific objects does only involve the field ϕ^i, whereas the abstract object recognition is supported by features ψ_i.

The learning of ψ_i is based on a formulation that closely resembles what has been done for ϕ^i, for which we have already considered the regularization issues. In the case of ψ_i, we can get rid of the trivial constant solution by minimizing[50]

$$I_\psi = \sum_{i=1}^{n} \int_\Gamma (1 - \psi_i(x, t))\phi^i(x, t) \, d\mu(x, t), \tag{3.23}$$

which comes from the p-norm translation of $\phi \overset{c}{\to} \psi$. Here, we are assuming that ϕ^i, ψ_i range in $[0, 1]$, so as whenever ϕ^i gets close to 1, it forces the same for ψ_i. This yields a well-posed formulation, thus avoiding the trivial solution.

The inherent affordance ψ_i that we characterize with Eq. (3.22), as we already observed, is tightly related to the feature field ϕ^i; indeed, they share the same index i and the velocity field v^i characterizes the development of ψ_i in an essential way.

In order to abstract the notion of affordance even further, we can, for instance, proceed as follows: For each $k = 1, \ldots, n$, we can consider another set of fields $\chi_k \colon \Gamma \to \mathbb{R}$, each of which satisfies the following condition

$$\chi_k \bowtie v^j = 0, \quad j = 1, \ldots, n, \quad j \neq k. \tag{3.24}$$

---

[50] Regularization from logic implication.

In this way, the variables χ_k do not depend on a particular v_i but they will need to take into account, during their development the multiple motion fields at the same time.[51]

In order to avoid the trivial constant solution, like for ψ we ask for the minimization of the logic implication term

$$I_\chi = \sum_{\kappa=1}^{n} \int_\Gamma (1 - \chi_\kappa(x, t))\psi_\kappa(x, t)\, d\mu(x, t). \tag{3.25}$$

This comes from the p-norm translation of $\psi \xrightarrow{c} \chi$. While this regularization term settles the value of χ_κ on the corresponding ψ_κ, notice that the motion invariance condition (3.24) does not assume any privilege with respect to the *firing feature* ψ_κ. Unlike for ψ, the motion invariance on features χ is directly driven by coupled features only, which contributes to elevate abstraction and lose the link with its firing feature.[52]

Once the set of the χ_k is given, a method to select the most relevant affordances could be simply done by a linear combination. In other words, a subselection of $\chi_1, \ldots, \chi_n$ can be done by considering for each $l = 1, \ldots, n_\chi < n$ the linear combinations

$$X_l := \sum_{k=1}^{n} a_{lk}\chi_k, \tag{3.26}$$

where $A := (a_{lk}) \in \mathbb{R}^{n_\chi \times n}$ is a matrix of learnable parameters. See note that since $X_l \in \text{span}(\chi_1, \ldots, \chi_n)$, as we remarked in Sect. 3.4 it satisfies $X_l \bowtie v^j = 0$ for all $j = 1, \ldots, n$. We can also extend the neural-like computation stated by Eq. (3.8) so as to gain feature abstraction and reduction. The corresponding feature is denoted by $\vdash \chi$. It is worth mentioning that the learning of coefficients a_{lk} does not involve motion invariance principles. Interestingly, as stated in Sect. 6.1, they can be used for additional developmental steps like that of object recognition. For example, they can be learned under the classic supervised framework along with correspondent regularization.

Why a vision field theory?

Tab. 3.1 gives a summary of the vision field theory, which reminds us the role of the *weak*, *strong*, and *reduction* fields along with their interactions and regularization principles. We find it interesting to point out that the I and II Principles are independent of the visual agent, and they naturally emerge around the conjugation operator $\bowtie$, which expresses feature invariance with respect to motion. The richness of the visual information, the body of the agent, and the number of visual features come in place whenever we consider the specific visual agent which is involved in the learning process. Of course, rich visual sources do require a lot of visual and

---

[51] Logic regularization.

[52] Reduction and abstraction.

Table 3.1 Vision field theory

	Self	Weak	Strong	Reduction
VF	ϕ	ψ	χ	$\vdash \phi, \vdash \psi, \vdash \chi$
I, II Pr.	$\phi^i \bowtie v^i = 0$	$\psi_i \bowtie (v^i - v^j) = 0$	$\chi_i \bowtie v_\kappa = 0$	$\vdash u = \sum u$
Reg.	min E, Eq. (3.11)), min R, Eq. (3.13)	$\varphi \xrightarrow{c} \psi$, Eq. (3.23)	$\psi \xrightarrow{c} \chi$	min $\sum_\kappa a_\kappa^2$

Summary of the I and the II Principles of visual perception. The self, weak, and strong interactions obey conjugation equations derived from those principles. The reduced fields, denoted by $\vdash$, produce features with increasing degree of abstractness

reduction fields, which are also likely playing a prominent role in those cases. The presence of many features requires a correspondent computational structure that, in this book, is supposed to be a neural architecture. In the next chapter, the role of such a neural architecture is in fact discussed in detail and the connections with biology are explored. Interestingly, it is only when we involve the neural architecture that we begin appreciating the crucial role of FOA, which is even better disclosed as we think of the learning process.

The nature of the vision fields leads to intriguing connections with electromagnetism that emerge when restricting to the information-based principles dictated by vision fields. Let us consider the very interesting interpretation on the emergence of the E-B fields given by Richard Feynman in his lectures of physics [32]. In order to explain how cavity resonators work, he offers a nice view on the E-B interactions by beginning from the capacitor of Fig. 3.4. He pointed out that there is in fact a nice picture behind the emergence of the fields which comes from the mutual successive refinement that one can clearly see beginning from DC and then move to a signal with a certain frequency. At the beginning, there is a spatially uniform electric field, whereas the magnetic field is absent. As we move to a signal which changes over time the displacement current $\epsilon \partial_t E$ generates a magnetic field with circular symmetry. Since it changes itself with time, it subsequently generates an electric field and, therefore, we end up into a cyclic mutual interaction between E and B. Interestingly,

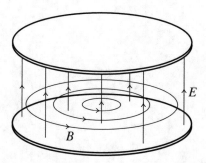

Fig. 3.4 Richard Feynman's view on the electromagnetic field as a successive refinement process to solve Maxwell's equations in a capacitor. The magnetic field arises with a circular structure as we feed the capacitor with AC. This, in turn, leads to generating the induced E fields

this way of interpreting the interaction leads to the correct solution, where the electric field, at a certain frequency becomes null on the border of the capacitor. In particular, when the applied signal has a frequency defined by $\omega = 2\pi f$, then we have[53]

$$E = E_o e^{j\omega t} J_o \left(\frac{\omega r}{c} \right),$$

where r is the radius of the plates, c is the speed of light, and J_o is the Bessel function. This leads us to promptly understand how a cavity resonator can be constructed from the capacitor. In Feynman's words: "... we take a thin metal sheet and cut a strip just wide enough to fit between the plates of the capacitor. Then we bend it into a cylinder with no electric field there ..."

This successive refinement scheme used for capturing the E-B electromagnetic interaction can be used also for vision fields. In our case, the field interaction is governed by the conjugation operator $\bowtie$. Like for electromagnetism, there is in fact an external environmental interaction, the video signal, which fuels the fields. The analogy is maintained when considering the case of DC in the capacitor. This corresponds with a spatiotemporal constant video that leads to constant features (E field) and null velocity (B field). As the video information comes, thanks to the presence of motion (the signal in the capacitor is not constant anymore), the change of the features allows us to track their change by developing the corresponding velocity.[54] Of course, that motion enables a corresponding adaptation of the visual fields and, like for the capacitor, we end up into a cyclic process that, hopefully, converges. We conjecture that this concretely happens whenever the video source presents appropriate statistical regularities. The other side of the medal is that of thinking of an upper bound on the corresponding amount of information that is associated with the source.

---

[53] Feynman's understanding on the birth of the $E - B$ electromagnetic field in a capacitor.

[54] Think of the classic brightness invariance or to the I Principle.

Chapter 4
Foveated Neural Networks

*The remarkable properties of some recent computer algorithms
for neural networks seemed to promise a fresh approach to
understanding the computational perspectives of the brain.
Unfortunately most of these neural nets are unrealistic in
important respects.*

Francis Crick, 1989

Keywords NN architectures · Foveated neural networks

4.1 Introduction

This chapter is about visual bodies, regardless of biology.[1] More than thirty years
ago, Francis Crick contributed to activate a discussion on the biological plausibility
of artificial neural nets along with the corresponding learning algorithms. As yet, this
is still an active research direction carried out in a battlefield that is full of slippery
and controversial issues. In the same paper of the above quotation [25], the author
clearly states that "a successful piece of engineering is a machine which does something useful. Understanding the brain, on the other hand, is a scientific problem."
He noticed that this scientific problem is a sort of reverse engineering on the prod-
uct of an alien technology, and it is the task of trying to unscramble what someone
else has made. The above statements can hardly be questioned. Biological processes
cannot neglect the importance of effectively sustaining a wide range of simultaneous tasks connected with the essence of life. However, the impressive development
of deep learning in the field of computer vision clearly indicates that there are in
fact artificial intelligence processes that make it possible to achieve concrete performance, especially relying on a the truly different computational structure. The
massive supervised learning protocol, along with the backpropagation algorithm, is
at the basis of spectacular results that could not be figured out at the dawn of the
connectionist wave mostly carried out by the parallel distributed processing (PDP)

---

[1] The slippery topic of biological plausibility.

A. Betti et al., *Deep Learning to See*, SpringerBriefs in Computer Science,
https://doi.org/10.1007/978-3-030-90987-1_4

research group [46, 47]. Interestingly, the introduction of the connectionist models and the development of deep learning [62] are not only having a great technological impact, but also opening the doors to the conception of computational processes that hold regardless of biology.[2] Considering the information-based principles and the vision field arguments on motion invariance discussed so far, in this chapter we claim that the computational mechanisms behind motion invariance can naturally be implemented by deep architectures that, throughout this book, are referred to as *foveated-based neural networks* (FNNs). They can in fact naturally carry out the conjugation of α_φ and α_v (see Eq. 3.9). This is expected to give rise to a machine capable of extracting codes that are motion invariant and that turn out to be very useful for visual perceptual tasks. While deep learning has mostly been sustained by the supervised learning protocol in computer vision, here deep architectures are only exposed to visual information provided by motion, with no labeled data. In this perspective, FNN shares with humans the same learning protocol and, hopefully, can also shed light on brain mechanisms of vision. The information-based principles that are presented hold regardless of biology. This chapter is organized as follows: In the next section, we address the crucial role of hierarchical architectures along with the assumption of using receptive fields for neural computation. In Sect. 4.3, we discuss the different roles of features charged of detecting "what" and "where" is in the visual scene and establish some biologic connection. In Sect. 4.4, we discuss variable resolution neural nets along with the relationship with classic convolutional nets and we address the overall neural architecture.

4.2 Why Receptive Fields and Hierarchical Architectures?

Beginning from early studies on the visual structure of the cortex [49], neuroscientists have gradually gained evidence on the fact that it presents a hierarchical structure and that neurons process the video information on the basis of inputs restricted to receptive fields. Interestingly, the recent spectacular results of convolutional neural networks suggest that hierarchical structures based on neural computation with receptive fields play a fundamental role also in artificial neural networks [62]. As stated in question 5 (**Q5**, Sect. 1.6), we may wonder why are there visual mainstreams organized according to a hierarchical architecture with receptive fields and if there is any reason why this solution has been developed in biology. First of all, we can promptly realize that, even though neurons are restricted to compute over receptive fields, deep structures rely on large virtual contexts for their decision. As we increase the depth of the neural network, the consequent pyramidal dependence that is established by the receptive fields increases the virtual input window used for the decision, so higher abstraction is progressively gained as we move toward the output. Hence, while one gives up on exploiting all the information available at a certain layer, the restriction to receptive field does not prevent from considering large

---

[2] Foveated-based neural networks.

windows for the decision. The marriage of receptive fields with deep nets turns out to be an important ingredient for a parsimonious and efficient implementation of both biological and artificial networks. In convolutional neural networks, the assumption of using receptive fields comes with the related hypothesis of *weight sharing* on units that are supposed to extract the same feature, regardless of where the neurons are centered in the retina. In so doing, we enforce the extraction of the same features across the retina. The same visual clues are clearly positioned everywhere in the retina, and the equality constraints on the weights turn out to be a precise statement for implementing a sort of *equivariance under translation*.

Clearly, this constraint has effect on invariance neither under scale nor under rotation. Any other form of invariance that is connected with deformations is clearly missed and is supposed to be learned. The current technology of convolutional neural networks in computer vision typically gains these invariances thanks to the power of supervised learning by "brute force." Notice that since most of the tasks involve object recognition in a certain environment, the associated limited amount of visual information allows us to go beyond the principle of extracting visual features at pixel level. Visual features can be shared over small windows in the retina by the process of pooling, thus limiting the dimension of the network. Basically, the number of features to be involved has to be simply related to the task at hand, and we can go beyond the association of the features with the pixels. However, the conquering of human-like visual skills is not compatible with this kind of simplifications since, as stated in the previous section, humans can perform pixel semantic labeling. There is a corresponding trend in computer vision where convolutional nets are designed to keep the connection with each pixel in the retina at any layer so as to carry out segmentation and semantic pixel label. Interestingly, this is where we need to face a grand challenge. So far, very good results have been made possible by relying on massive labeling of collections of images. While image labeling for object classification is a boring task, human pixel labeling (segmentation) is even worst! Instead of massive supervised labeling, one could realize that motion and focus of attention can be massively exploited to learn the visual features mostly in an unsupervised way. A recent study in this direction is given in [9], where the authors provide evidence of the fact that receptive fields do favor the acquisition of motion invariance which, as already stated, is the fundamental invariance of vision. The study of motion invariance leads to dispute the effectiveness and the biological plausibility of convolutional networks. First, while weight sharing allows us to gain translational equivariance on single neurons, the vice versa does not hold true. We can think of receptive field-based neurons organized in a hierarchical architecture that carry out translation equivariance without sharing their weights. This is strongly motivated also by the arguable biological plausibility of the mechanism of weight sharing [75]. Such a lack of plausibility is more serious than the supposed lack of a truly local computational scheme in backpropagation, which mostly comes from the lack of delay in the forward model of the neurons [13].

Hierarchical architectures are the natural solution for matching the notion of receptive field and develop abstract representations. This also seem to facilitate the implementation of motion invariance, a property that is at the basis of the biological structure of the visual cortex. The architectural incorporation of this fundamental invariance property, as well as the match with the need for implementing the focus of attention mechanisms, likely needs neural architectures that are more sophisticated than current convolutional neural networks. In particular, neurons which provide motion invariance likely benefit from dropping the weight sharing constraint.

4.3 Why Two Different Mainstreams?

In Sect. 1.4, we have emphasized the importance of bearing in mind the neat functional distinction between vision for action and vision for perception. A number of studies in neuroscience lead to conclude that the visual cortex of humans and other primates is composed of two main information pathways that are referred to as the ventral stream and dorsal stream [40, 41]. We typically refer to the ventral "what" and the dorsal "where/how" visual pathways. The ventral stream is devoted to perceptual analysis of the visual input, such as object recognition, whereas the dorsal stream is concerned with providing spatial localization and motion ability in the interaction with the environment. The ventral stream has strong connections to the medial temporal lobe (which stores long-term memories), the limbic system (which controls emotions), and the dorsal stream. The dorsal stream stretches from the primary visual cortex (V1) in the occipital lobe forward into the parietal lobe. It is interconnected with the parallel ventral stream which runs downward from V1 into the temporal lobe. We might wonder: Why are there two different mainstreams? What are the reasons for such a different neural evolution? (see **Q6**, Sect. 1.6). This neurobiological distinction arises for effectively facing visual tasks that are very different. The exhibition of perceptual analysis and object recognition clearly requires computational mechanisms that are different with respect to those required for estimating the scale and the spatial position. Object recognition requires the ability of developing strong invariant properties that mostly characterize the objects themselves. By and large, scientists agree on that objects must be recognized independently of their position in the retina, scale, and orientation. While we subscribe this point of view, a more careful analysis of our perceptual capabilities indicates that these desirable features are likely more adequate to understand the computational mechanisms behind the perception of rigid objects. The elastic deformation and the projection into the retina give in fact rise to remarkably more complex patterns that can hardly be interpreted in the framework of geometrical invariances. We reinforce the claim that *motion invariance is in fact the only invariance which does matter*. Related studies in this direction can be found [8]. As the nose of a teddy bear approaches children's eyes, it becomes larger and larger. Hence, scale invariance is just a by-product of motion invariance. The same holds true for rotation invariance. Interestingly, as the children deforms the teddy bear a new visual pattern is created that, in any case, is the outcome of the motion of "single object particles." The neural enforcement of motion invariance is conceived by implementing the "what" neurons. Of course, neurons with built-in motion invariance are not adequate to make spatial estimations or detection of scale/rotation. Unlike the "what" neurons, in this case motion does matter and the neural response must be affected by the movement.

These analyses are consistent with neuroanatomical evidence and suggest that "what" and "where" neurons are important in machines too. The anatomical difference of the two mainstreams is in fact the outcome of a truly different functional role. While one can ignore such a difference and rely on the rich representational power of big deep networks, the underlined difference stimulates the curiosity of discovering canonical neural structures to naturally incorporate motion invariance, with the final purpose of discovering different features for perception and action. The emergence of the indissoluble pair (φ, v_φ) helps understanding the emergence of neurons specialized on the different functions of perception and action.

4.4 Foveated Nets and Variable Resolution

The implementation of motion invariance based on conjugate features does require the involvement of the specific structure of φ and v_φ.[3] Let us assume that they come from neural networks that carry out computations on pixel x at time t of the given video signal. Of course, the neural networks are expected to take a decision at t that depends on the entire frame and on the specific pixel of coordinate x where we want to compute the features. As already pointed out, following Eq. (3.9), we can consider deep neural networks based on receptive fields, which means that we can always regard the output of neurons as functions of the corresponding virtual window. Let us focus on the computation of features φ. In that case, suppose $\Omega_\varphi(x)$ is the virtual window corresponding with φ that is centered on x. Two deep neural networks can be used to express functions α_φ and α_v. This leads to adopt the following computational model [compare what follows with Eq. (3.9)]:

$$
\begin{aligned}
\varphi(x, t) &= \alpha_\varphi(b(\cdot, t), a(t), t)(x) = \eta_\varphi\big(w_\varphi(\cdot, t, a(t)), b(\cdot, t)\big)(x) \\
v_\varphi(x, t) &= \alpha_v(D\varphi(\cdot, t, a(t), t)(x) = \eta_v\big(w_{v_\varphi}(\cdot, t, a(t)), \nabla\varphi(\cdot, t)\big)(x),
\end{aligned}
\tag{4.1}
$$

---

[3] Given a video, for conjugate features to exist we need an appropriate computational model, which corresponds with expressing the structure of the pair (φ, v).

Of course, the conjugation $\varphi \bowtie v_\varphi$ does involve the choice of both $\eta_\varphi \colon \mathbb{R}^N \times \mathbb{R}^\Omega \to \mathbb{R}^\Omega$ and $\eta_v \colon \mathbb{R}^M \times \mathbb{R}^\Omega \to \mathbb{R}^\Omega$, along with their weights $w_\varphi \colon \Gamma \times \Omega \to \mathbb{R}^N$ and $w_{v_\varphi} \colon \Gamma \times \Omega \to \mathbb{R}^M$ that are expected to depend on the pixel on the retina and on the point of focalization of attention around which the foveated structure, as we will see in what follows, is centered. This is a remarkable difference with respect to convolutional networks, where one assumes the weight sharing of the weights. For CNN, the weight sharing is a fundamental constraint to make the feature extraction independent of the pixel. However, the assumption of acquiring visual frames under the driving mechanisms of focus of attention changes the cards on the table. [4] As pointed out in the list of motivations for choosing foveated eyes in item (2) of Sect. 2.3, there are good reasons for choosing variable resolution when considering the temporal structure of the computation and the focus of attention mechanisms. Moreover, still in the same list, the claim in (2) puts forward the limitations of weight sharing when considering the need of disambiguating decisions in different positions in the retina. Interestingly, as we assume higher resolution close to the focus of attention we somewhat face the issue reported in item (2). In the following, we discuss the impact of variable retina resolution, which is strongly connected with FOA, in the architecture of neural network η_φ. Of course, the same arguments hold for neural networks η_v.

Now let us regard the weights ω_φ^x of the single neurons of η_φ. For every fixed point on the retina $x \in \Omega$, and for every temporal instant $t \in [0, T]$ let $\omega_\varphi^x(\cdot, t) \colon \mathbb{R}^2 \to \mathbb{R}$ be a compactly supported filter such that

- $0 \in \mathrm{supp}(\omega_\varphi^x(\cdot, t)) \subset \mathbb{R}^2$,
- $\mathrm{diam}\,\mathrm{supp}(\omega_\varphi^x(\cdot, t)) < \mathrm{diam}\,\Omega$,

where $\mathrm{diam}\,A := \sup\{|x - y| : x, y \in A\}$ and supp is the support of the filter. In the continuum setting, those filters are used to compute the activation v of any neuron of η_φ as

$$
\begin{cases}
v_\varphi(x, t) = (\omega_\varphi^x(\cdot, t) \star \hat{y}(\cdot, t))(x) := \displaystyle\int_{\mathbb{R}^2} \omega_\varphi^x(x - \xi, t)\hat{y}(\xi, t)\,d\xi, \\[2mm]
\varphi(x, t) = \gamma(v_\varphi(x, t))
\end{cases}
\tag{4.2}
$$

where $y \colon \Omega \times [0, T] \to \mathbb{R}$ is the output of the previous layer,[5] and $\hat{y}(\cdot, t) \colon \mathbb{R}^2 \to \mathbb{R}$ is the extension of y to $\mathbb{R}^2$ with $\hat{y}(x, t) = 0$ for all $t \in [0, T]$ if $x \notin \Omega$ (zero padding). As usual, we assume that the output φ is computed from the activation v_φ by the neural output function γ.

---

[4] Variable resolution and weights depending on the position in the retina.

[5] Here for simplicity, we are considering the scalar case, although the extension to vectorial features is straightforward.

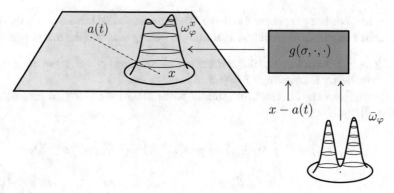

Fig. 4.1 The computation of the filters ω_φ^x is done through the function g once the distance from the focus of attention is established

Consider now the following way to compute the receptive field-based filters ω_φ^x:

$$\omega_\varphi^x(z, t) = \int_{\mathbb{R}^2} g(\sigma, x - a(t), z - \xi)\hat{\bar{\omega}}_\varphi(\xi, t) \, d\xi, \tag{4.3}$$

where $t \mapsto a(t)$ is a given trajectory (in our case it will be the trajectory Fig. 4.1 of the focus of attention), $\bar{\omega}_\varphi \colon K \times [0, T] \to \mathbb{R}$ is a single set of weights for each temporal instant defined over the compact set $K \subset \mathbb{R}^2$ containing the origin, and $g \colon \bar{\mathbb{R}}_+ \times \mathbb{R}^2 \times \mathbb{R}^2 \to \mathbb{R}$ is an appropriate filter which depends on the parameter σ and takes into account the position of the point $x \in \Omega$ with respect to the focus of attention. Clearly if $g(\sigma, d, \cdot)$ is compactly supported, if $0 \in \mathrm{supp}(g(\sigma, d, \cdot))$, and if $\mathrm{diam}\, K + \mathrm{diam}\,\mathrm{supp}(g(\sigma, d, \cdot)) < \mathrm{diam}\,\Omega$ for any $\sigma \in \bar{\mathbb{R}}_+$ and for every $d \in \mathbb{R}^2$, then the two above conditions are satisfied.

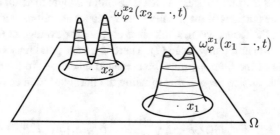

Now, let us analyze the role of the focusing function g. We consider two extreme cases that give insights on the role of g under the assumption that $g(\sigma, d, x) = G(|d|, x)$ where $G \colon \bar{\mathbb{R}}_+ \times \mathbb{R}^2 \to \mathbb{R}$ and such that $\lim_{\varepsilon \to 0} G(\varepsilon, x) = \delta$ (near the focus of attention the filter g is extremely sharp). We also assume that $\lim_{z \to \text{diam } \Omega} G(z, \cdot) = C$, i.e., far away from the focus of attention the filter g becomes flat and constant[6]

1. $|x - a(t)| \approx 0$: In this case under the above assumptions $g(\sigma, d, x) = \delta(x)$; then, it immediately follows from Eq. (4.3) that $\omega_\varphi^x \equiv \hat{\bar{\omega}}_\varphi$.
2. $|x - a(t)| \approx \text{diam } \Omega$: Here, we simply have, due to the fact that $\lim_{a \to \text{diam } \Omega} G(a, \cdot) = C$,

$$\omega_\varphi^x(z, t) = C \int_K \bar{\omega}_\varphi(\xi, t) \, d\xi =: C\mathcal{L}^2(K)\langle \bar{\omega}(\cdot, t)\rangle_K, \quad \forall z \in \mathbb{R}^2,$$

where $\mathcal{L}^2(K)$ is the Lebesgue measure of K. Plugging this expression back into Eq. (4.2), we get

$$v_\varphi(x, t) = C\mathcal{L}^2(K)\langle \bar{\omega}(\cdot, t)\rangle_K \int_{\mathbb{R}^2} \hat{y}(\xi, t) \, d\xi =: C\mathcal{L}^2(K)\langle \bar{\omega}(\cdot, t)\rangle_K \langle y(\cdot, t)\rangle_\Omega,$$

which means that when we are away from the focus of attention the value of the activations is spatially constant which is what we expect from a low-resolution signal.

It turns out that FNN is based on filters that are somewhat in between those associated with these two extreme cases. Roughly speaking, this forces the learning process to be very effective where the fovea is focusing attention, whereas as we move far away, the learning process only acts on possible modifications of the average of the weights of the filter. This somewhat reflects the downsampling that arises on those pixels where only the spatial average of the signal on the receptive field is actually perceived. It is worth mentioning that the removal of the weight sharing principle leads to promote a learning process which is stimulated more to adapt the parameters on the basis of the portions of the retina where we focus attention. In any pixel which is far away from the focus, we only take into account the average of the weights, so as to better face the ambiguity issue (7). There is a very good reason for assuming that the weights of the filter $\bar{\omega}_\varphi$ must be averaged when we are far away from the FOA.[7] A possible choice for g is that of using the family of truncated Gaussians[8]

$$g(\sigma, d, x) := \frac{1}{\sqrt{2\pi}(\sigma + |d|)} \exp\left(-\frac{x^2}{2(\sigma + |d|)^2}\right) 1_{\{|x| < \rho\}}(x). \qquad (4.4)$$

---

[6] This assumption is in contradiction with the fact that g should have compact support; however, it saves us from taking into account boundary effects that would raise technical issues that in our opinion would go against the purpose of giving clear insight on the role of g.

[7] Gaussian focusing filter.

[8] Remember that g has to have compact support.

Fig. 4.2 The computation of the filter g can be in general carried out by a neural network with parameters σ. Such network at each temporal instance t should also take as input the displacement $x - a(t)$ of the point x in which we want to compute the feature φ from the focus of attention $a(t)$

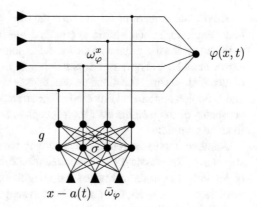

Here, ρ represent the radius at which we cut the tails of the Gaussian to zero. If we regard σ as a learnable parameter which depends on t, then the learning process could begin with very high level of σ and decrease to values close to zero. Hence, at the beginning of learning we have $\sigma \to \infty$ which yields null weights. As time goes by, the value of σ is gradually reduced until $g \to 0$ which, as already stated, yields around the focus of attention the maximum visual acuity stated by $\omega_\varphi^x \equiv \hat{\omega}_\varphi$.

In the above discussion, we always regarded the parameter σ that defines the family of maps $(d, x) \mapsto g(\sigma, d, x)$ as a real number; in general, however, it could be extremely interesting in order to augment the expressivity of the filters ω_φ^x to consider the case in which $(d, x) \mapsto g(\sigma, d, x)$ is indeed a neural network and σ are its weights (as shown in Fig. 4.2).

Circular Crown FNN

We can approximate the above computational model by defining receptive field base filters whose shape changes depending on where we focus attention. Let $R :=$ diam Ω. Suppose also that we are given a collection of m filters $\omega_1, \ldots, \omega_m$ each of which, for each fixed temporal instant, is a map $\omega_i(\cdot, t) \colon \mathbb{R}^2 \to \mathbb{R}$ with compact support which contains the origin.

Then fixed $t \in [0, T]$, and given the focus of attention trajectory $t \mapsto a(t)$, we define for all $x \in \Omega$

$$
\omega^x(\xi, t) := \begin{cases} \omega_1(\xi, t) & |x - a(t)| \leq R/m; \\ \omega_2(\xi, t) & R/m < |x - a(t)| \leq 2R/m; \\ \vdots & \vdots \\ \omega_m(\xi, t) & (m-1)/m < |x - a(t)| \leq R; \end{cases}
$$

Notice indeed that there cannot be a $x \in \Omega$ that for some $t \in [0, T]$ yields $|x - a(t)| > R$ by definition of diameter of Ω.

Additional constraint on the form of the filters $\omega_1, \ldots, \omega_m$ can be imposed to better exploit the focalization of attention mechanism; for instance, it is reasonable to think that the filter that acts closer to the point where we focus attention should be narrower and could have rapid spatial variations in order to allow for crisper transfer of information while the filters whose index is closer to m should probably be broader and quite flat. Notice that this kind of constraints could be implemented by imposing constancy of the weights over their receptive fields when we move farther from the focus of attention.

Another important remark is in order; when we choose $\omega_1 \equiv \omega_2 \equiv \cdots \equiv \omega_m$ we recover the classical CNN architecture. This fact suggests that at the beginning of learning the initialization of the weights could be done uniformly $\omega_1(\cdot, 0) = \omega_2(\cdot, 0) = \cdots = \omega_m(\cdot, 0) = \omega^0$ at least to suggest to the network an initial equivariance under translation.

Propagation of the Visual Signal Through the Network

The propagation of visual information that has been assumed in Eq. (4.2) is the typical forward *instantaneous* propagation of NN. This basically means that in Eq. (4.2) we are assuming that the speed of propagation of the signal from the input to the output of the neural network is much greater than the typical dynamic of the video signal.

In general however, in order to avoid some conceptual paradoxes (see [14]), it is much more natural to consider a propagation dynamic from input to output. One of the simplest ways to model this propagation is to substitute Eq. (4.2) with

$$\begin{cases} \upsilon_\varphi(x, t) = (\omega_\varphi^x(\cdot, t) \star \hat{y}(\cdot, t))(x); \\ \varphi_t(x, t) = cF(\varphi(x, t), \gamma(\upsilon_\varphi(x, t))), \end{cases} \tag{4.5}$$

where $F \colon \mathbb{R} \times \mathbb{R} \to \mathbb{R}$ specifies the way in which the propagation takes place. For instance, a possible choice here would be $F(a, b) = a - b$; in this case, we would have

$$\varphi_t(x, t) = c[\varphi(x, t) - \sigma(\upsilon_\varphi(x, t))].$$

This form makes also clear the role of the propagation velocity c: When c is big, then the above equation formally reduces to the static propagation described in Eq. (4.2).

It is also important to compare this newly introduced dynamics of the features with the computational model that we introduced in Chap. 3, specifically in Eq. (3.9). This two dynamics are in fact rather different: The dynamics proposed in Eq. (3.9) was an added dynamic that was restricted at the level of the feature φ, and its computation still involved a classical forward computation in order to determine the value of α_φ, while in Eq. (4.5) we are introducing a temporal delay in the computation of the features at a certain level of our architecture in terms of the value of the features at previous layers.

Chapter 5
Information-Based Laws of Feature Learning

When I was in high school, my physics teacher—whose name was Mr. Bader—called me down one day after physics class and said, "you look bored; I want to tell you something interesting." Then he told me something which I found absolutely fascinating, and have, since then, always found fascinating. Every time the subject comes up, I work on it.

Richard Feynman, physics lectures about the principle of least action

Keywords Optical flow estimation · NN representation · Online learning · Optimal control

5.1 The Simplest Case of Feature Conjugation

The brightness is the simplest feature which only involves a single pixel. Hence, we are in front of the simplest case of feature conjugation $b \bowtie v = 0$ that has been discussed in Sect. 3.4.[1] The well-posedness of the problem that is gained by the constrained minimization of the functional defined by Eq. 3.2 is typically given a different formulation, which only involves a soft satisfaction of $b \bowtie v = 0$. This can be formulated by stating that, given b, one must minimize

$$E(t) = \int_{\Omega} (b \bowtie v(x, t))^2 + \lambda\big((\nabla v_1(x, t))^2 + (\nabla v_2(x, t))^2\big)\, dx, \qquad (5.1)$$

where $\lambda > 0$ is the regularization parameter. The soft enforcement of the conjugation constraint in this case makes sense. As already pointed out, this conjunction arises from the brightness invariance principle, which is only an approximate description of what happens in real video. The stationary point of E, which is also a minimum, is obtained by using the Euler–Lagrange equations

---

[1] A soft satisfaction of the $b \bowtie v = 0$ conjugation.

© The Author(s), under exclusive license to Springer Nature Switzerland AG 2022
A. Betti et al., *Deep Learning to See*, SpringerBriefs in Computer Science,
https://doi.org/10.1007/978-3-030-90987-1_5

$$\begin{cases} \nabla^2 v_1 = \frac{1}{\lambda}(b \bowtie v)\, \partial_{x_1} b \\ \nabla^2 v_2 = \frac{1}{\lambda}(b \bowtie v)\, \partial_{x_2} b, \end{cases} \tag{5.2}$$

that are paired with appropriate boundary conditions on v.[2]

The regularization term (3.2) helps facing the ill-posedness of brightness invariance. Basically, one must distinguish the motion of material points and their project onto the retina and the corresponding optical flow that is interpreted as it is in an inherently ambiguous context.

While the regularization helps discovering one precise solution, it might not be really satisfactory in terms of the visual interpretation of the environment. Something similar happens in the brain. Many motion sensitive neurons in the visual system close to the retina respond locally while processing on small receptive fields. In higher level of the hierarchical architecture of the visual system, the information from several of the neurons that respond locally is integrated to allow us to perceive more global information. The inherent ambiguity in motion computation in neurons based on a small receptive field is called the *aperture problem*. The aperture problem arises when looking at a moving image through a small hole—the aperture. Different directions of motion can appear identical when viewed through an aperture[3]. In order to conquer a deeper interpretation, we need to extract higher-level features along with their own conjugated velocity.

5.2   Neural Network Representation of the Velocity Field

Now consider the case in which the velocity field is estimated through a neural network $\eta_v \colon \mathbb{W}_x \times C^\infty(\Omega; \mathbb{R}^{3c}) \to (\mathbb{V}_x)^2$ where c is the number of channels of the brightness. Here, $\mathbb{W}_x$ is the functional space that determines the spatial regularity of the filters, while $\mathbb{V}_x$ is the space in which we select the velocity field at each temporal instant as a function of space; for the moment however, we do not want to discuss what could be appropriate choices of those spaces. This map uses a set of filters $\omega \colon \Omega \to \mathbb{R}^N$ to associate with the spatiotemporal gradient of the brightness $Db(\cdot, t)$ evaluated at a certain temporal instant t to the corresponding velocity field $v(\cdot, t) = \eta_v(\omega, Db(\cdot, t))$.

Notice that at this level the "neural network" η_v is a potentially very complicated map between functional spaces, and thus, it is something that would be rather delicate to work with. For this reason, we abandon the very useful picture of the theory on a continuous retina, and for the remainder of this section, we will focus on a discrete description of the retina made up of discrete entities that we will denote as pixels. However, we will still retain continuity in the temporal domain since this will allow

---

[2] The aperture problem and the barber's pole.

[3] A very nice example is illustrated at http://elvers.us/perception/aperture/.

us to describe the learning process in terms of a continuous temporal dynamics that in our opinion is very helpful.

So, we model the retina as the set $\Omega^\diamond := \{(i, j) \in \mathbb{N}^2 : 1 \leq i \leq w, 1 \leq j \leq h\}$. Then, the brightness map can be modeled as a map $t \mapsto b(t) \in \mathbb{R}^{w \times h}$.[4] We will also denote with $t \mapsto b'(t) \in \mathbb{R}^{w \times h}$ the discretization of the temporal partial derivative $b_t(x, t)$ of the brightness. The discrete spatial gradient operator $D: \mathbb{R}^{w \times h} \to \mathbb{R}^{w \times h \times 2}$ defined by its action on any $x \in \mathbb{R}^{w \times h}$

$$(Dx)_{i,j,1} = \begin{cases} x_{i+1,j} - x_{i,j} & \text{if } 1 \leq i < w; \\ 0 & \text{otherwise,} \end{cases}$$

$$(Dx)_{i,j,2} = \begin{cases} x_{i,j+1} - x_{i,j} & \text{if } 1 \leq j < h; \\ 0 & \text{otherwise.} \end{cases}$$

the network that estimates discrete, the a network that estimates the velocity field f should be in general a function that given some set of N parameters, a spatial discrete gradient of an image, and its temporal partial derivative computes a two dimensional vector (the optical flow) for each pixel of the retina. More formally given the function $f: [0, T] \times \mathbb{R}^N \times \mathbb{R}^{w \times h \times 2} \times \mathbb{R}^{w \times h} \to \mathbb{R}^{w \times h \times 2}$, we are after the weight maps $t \mapsto w(t) \in \mathbb{R}^N$ such that $f(t, w(t), Db(t), b'(t))$ should as times go by be a good estimation of the optical flow. Here, the extra temporal dependence of the neural network is added since we also want to take into account in the computation the value of $a(t)$, i.e., of the position of the focus of attention.

In a continuous setting like the one in which we pose ourselves, it is natural to consider an online learning problem for the parameters w. A very natural way in which this problem can be formalized is as a non-homogeneous (online) flow on a suitably regularized Horn and Schunk functional. To pose this problem, define the map $(x, s) \in [0, T] \times \mathbb{R}^N \mapsto HS(x, s) \in \mathbb{R}_+$ where

---

[4] Here, for simplicity and for not having to deal with too many indices, we are considering images which have only one channel, like grayscale images; however, if we let $t \mapsto b(t) \in \mathbb{R}^{w \times h \times c}$ where c is the number of channels of the image at time t (e.g., $c = 3$ for RGB images), we recover the general case.

$$HS(x, s) := \frac{1}{2} \sum_{\substack{1 \le i \le w \\ 1 \le j \le h}} \Bigg(((Db(s))_{i,j,1}, (Db(s))_{i,j,2}, b'(s))$$

$$\cdot \begin{pmatrix} \begin{pmatrix} f_{i,j,1}(s, x, Db(s), b'(s)) \\ f_{i,j,2}(s, x, Db(s), b'(s)) \\ 1 \end{pmatrix} \end{pmatrix}^2 \Bigg)$$

$$+ \frac{\lambda}{2} \sum_{\substack{1 \le i \le w \\ 1 \le j \le h}} \Big((Df_{i,j,1}(s, x, Db(s), b'(s)))^2 $$

$$+ (Df_{i,j,2}(s, x, Db(s), b'(s)))^2 \Big),$$

(5.3)

is the explicit version of the Horn and Schunk functional regarded as a function of time (that as we will take the place of the parameter s) and the weights of the network (that will occupy the position taken by x). We will also denote with $\nabla HS(x, s) \equiv HS_x(x, s)$ the gradient with respect to the first argument of HS (i.e., the parameters of the network) and with $HS_t(x, s)$ the partial derivative with respect to the second argument (i.e., partial temporal derivative).

We want to define the learning dynamic as the outcome of a variational principle (known as the WIE principle [65]) on the functional

$$F_\varepsilon(\omega) := \int_0^T e^{-t/\varepsilon} \left(\frac{\nu\varepsilon}{2} |\omega'(t)|^2 + HS(\omega(t), t) \right) dt$$

(5.4)

defined over the set $\mathbb{X} = \{\omega \in H^1([0, T]; \mathbb{R}^N) : \omega(0) = \omega^0\}$. The wanted learning trajectory then it is obtained (see [11, 65]) as follows:

1. For each fixed ε, find a minimizer ω_ε of (5.4).
2. Take the limit $\varepsilon \to 0$ and consider the limit trajectory (in a suitable topology).

If we assume that ω is smooth[5] using Eqs. (A.5) and (A.6) we can write the stationarity condition for F_ε as

$$\begin{cases} -\varepsilon\nu\ddot{\omega}_\varepsilon(t) + \nu\dot{\omega}(t) + \nabla HS(\omega(t), t) = 0 & t \in (0, T) \\ \omega_\varepsilon(0) = \omega^0, \quad \varepsilon\nu\omega_\varepsilon(T) = 0. \end{cases}$$

(5.5)

Then, taking the formal limit as $\varepsilon \to 0$ yields the following dynamics:

$$\begin{cases} \omega'(t) = -\frac{1}{\nu}\nabla HS(\omega(t), t) \\ \omega(0) = \omega^0. \end{cases}$$

(5.6)

---

[5] This can actually be proven assuming regularity on $HS(\cdot, t)$; in particular, if $HS(\cdot, t) \in C^\infty(\mathbb{R}^N)$, then also $\omega \in C^\infty([0, T])$.

Notice that in order to compute HS, we should at some point compute the gradient of the network with respect to its parameters; this can be done algorithmically using the backprop.

Now notice that Eq. (5.6) represents a non-homogeneous flow on the "potential" HS, and the non-homogeneity comes from the fact that since we are dealing with a video stream the gradients of the brightness change all the time. The goal of the learning is to reach a set of constant weights that properly estimate the optical flow given any couple of frames in the video. We argue however that the learning process itself could benefit from a developmental stage in which the task is simplified by means of a spatial filtering of the input.

In order to do this, we consider the following "smoothing" process of the input achieved, for instance, by a Gaussian-like filtering; given a frame $x \in \mathbb{R}^{w \times h}$ and a parameter $\sigma \geq 0$, we consider

$$x \mapsto \Phi(x, \sigma) \in \mathbb{R}^{w \times h}.$$

We assume that this filtering should give back the frame for $\sigma = 0$, i.e., $\Phi(x, 0) = x$ and that for $\sigma > 0$ should satisfy $\|D\Phi(x, \sigma)\| \leq \|Dx\|$.

Now the idea is that the degree of filtering that is expressed by σ could be learned alongside with the weights of the network. Of course in order to avoid trivial solutions, we should also require that this smoothing should become as small as possible.

With this filtering of the input, our HS potential defined in Eq. (5.3) should take into account an additional dependence. This can be readily done by considering $\overline{HS}: [0, T] \times \mathbb{R}^N \times \mathbb{R} \to \mathbb{R}_+$ defined by[6]

$$
\begin{aligned}
\overline{HS}(s, x, \sigma) := \ &\frac{1}{2} \Bigg(\big((D\Phi(b(s), \sigma))_{i,j}, \Phi'(b(s), \sigma) \big) \\
&\cdot \left(\begin{array}{c} f_{i,j}(s, x, D\Phi(b(s), \sigma), \Phi'(b(s), \sigma)) \\ 1 \end{array} \right)^2 \Bigg) \\
&+ \frac{\lambda}{2} |D f_{i,j}(s, x, D\Phi(b(s), \sigma), \Phi'(b(s), \sigma))|^2,
\end{aligned}
\tag{5.7}
$$

where $\Phi'(b(s), \sigma)$ is the analogue of $b'(s)$ after the Φ transformation is applied.

Correspondingly, the functional in Eq. (5.4) becomes

$$
\overline{F}_\varepsilon(x, \sigma) := \int_0^T e^{-t/\varepsilon} \Bigg(\frac{\nu_\omega \varepsilon}{2} |\omega'(t)|^2 + \frac{\nu_\sigma \varepsilon}{2} (\sigma'(t))^2 \\
+ \overline{HS}(t, \omega(t), \sigma(t)) + \frac{k}{2} (\sigma(t))^2 \Bigg) dt.
$$

---

[6] Here, we use Einstein convention and we do not explicitly write the two components of f and $D\Phi$ for compactness.

Taking the same exact steps that led to Eq. (5.6) with the additional variable σ, we get

$$\begin{cases} \omega'(t) = -\frac{1}{\nu_\omega}\nabla\overline{HS}(t, \omega(t), \sigma(t)) \\ \sigma'(t) = -k\sigma(t) - \frac{1}{\nu_\sigma}\overline{HS}_\sigma(t, \omega(t), \sigma(t)) \\ \omega(0) = \omega^0, \quad \sigma(0) = \sigma^0. \end{cases} \tag{5.8}$$

Here, $\sigma^0 > 0$ is a parameter chosen in such a way that $\Phi(b(0), \sigma^0)$ is a constant (or constant up to a certain tolerance) frame.

5.3 A Dynamic Model for Conjugate Features and Velocities

In the previous section, we described an online method to extract the optical flow from the brightness.

We now turn to the definition of an online method for the learning of the features and of their conjugate velocities as it is described in Chap. 3. In order to do this, we need to lay down some useful notation. As we did in the previous section, we will still work on the discrete retina Ω°.

The first ingredient that we need to put in the theory, as it is widely discussed in Chap. 3, is a set of features that are instrumental to the identification of objects. Here, we consider a set of maps $P^1, \ldots, P^m$ such that $P^k : \mathbb{R}^{w\times h\times m_{k-1}} \to \mathbb{R}^{w\times h\times m_k}$ with $m_0 = 1$ (or $m_0 = 3$ if we consider RGB images). Examples of these transformation could be convolution or, better yet, the foveated layer transformation described in the previous chapter. In any case, the assumption is that each of these maps can be regarded as parametric maps $\Pi^k : \mathbb{R}^{N_k} \times \mathbb{R}^{w\times h\times m_{k-1}} \to \mathbb{R}^{w\times h\times m_k}$ so that $\Pi^k(\omega^k, \cdot) \equiv P^k(\cdot)$ for some $\omega^k \in \mathbb{R}^{N_k}$, $k = 1, \ldots, n$. Once the dynamics of the set of weights $t \mapsto \omega^k(t)$ has been fixed, the resulting maps obtained from the composition of the Π^k are exactly the feature maps on which we based our discussion in the previous chapters:

$$\varphi^i(t) := \left(\bigcirc_{k=1}^{i} \Pi^k(\omega^k(t), \cdot) \right)(b(t))$$

where $\circ$ means function composition and where as usual here with φ^i we are thinking about the discretized version of the feature maps φ^i defined on the cylinder Γ. Consistently, as we have done in the previous section however we still retain the continuous temporal dependence.

However, since these are the building blocks of our theory, and we need to prescribe a dynamics for the weights, it is also convenient to explicitly consider the dependence of the features on the weights. To do this, we instead define the set of maps $A^i : \prod_{k=1}^{n} \mathbb{R}^{N_k} \times [0, T] \to \mathbb{R}^{w\times h\times m_i}$ defined as

$$A^i(\omega, t) := \left(\bigcirc_{k=1}^i \Pi^k(\omega^k, \cdot) \right)(b(t))$$

where in this case ω^k is thought as the kth component of ω in the space $\prod_{k=1}^n \mathbb{R}^{N_k}$ (i.e., $\omega = (\omega^1, \omega^2, \ldots, \omega^n)$). It is important to notice that at this stage we can think of the features φ^i either as scalar features as it has been done till now (in that case $m_k \equiv 1$ for all $k = 1, \ldots, n$) or as vector which codes a single feature (like it happens for RGB signals).

The same exact construction can be done for the affordance features ψ^i and χ^i, and yields the map B for the ψ features and the C map for the χ features.

Now, the learning process is therefore described by the dynamics of the weights $t \mapsto \omega_A(t)$, $t \mapsto \omega_B(t)$, and $t \mapsto \omega_C(t)$. As a note on notation, we will denote with $\nabla A^i : \prod_{k=1}^n \mathbb{R}^{N_k} \times [0, T] \to \mathbb{R}^{w \times h \times m_i} \times \prod_{k=1}^n \mathbb{R}^{N_k}$ the gradient of A^i with respect to its first argument (the parameters of the network ω) and with $A^i_t : \prod_{k=1}^n \mathbb{R}^{N_k} \times [0, T] \to \mathbb{R}^{w \times h \times m_i}$ the partial derivative with respect to its second argument (i.e., the partial derivative with respect to time). On the other hand, consistently with what we did in the previous section we assume that the spatial gradient operator D acts separately on each of the m_i components of the feature so that, for instance, for each admissible ω, t and i we have $DA^i(\omega, t) \in \mathbb{R}^{w \times h \times m_i \times 2}$. Similar definitions will be used for the maps B^i and C^i.

Moreover for each feature, φ we need to model, again with a neural network, the corresponding optical flow. As we have previously discussed, the input of this network will be the gradients of the brightness, i.e., Db and b_t. We will denote the parameters of the network that estimate the velocity as $\vartheta^k \in \mathbb{R}^{M_k}$, $v^i : \prod_{k=1}^n \mathbb{R}^{M_k} \times [0, T] \to \mathbb{R}^{w \times h \times 2}$. As it happened for the features here we compactly express the dependence of v^i on the whole set of parameters of the network that estimate the velocities; again the learning process should describe the trajectory $t \mapsto \vartheta(t)$.

The last ingredient that we need to lay down the learning problem is the "decoding" part of the architecture. As we described in Chap. 3, a natural regularization for the features is the reconstruction (or better yet the time shifted reconstruction); such map should take as input the features $\varphi^i(t)$ and decode them into the shifted frame $b(t + \tau)$[7]. So, we define the map $\Delta : \mathbb{R}^d \times \prod_{k=1}^n \mathbb{R}^{w \times h \times m_k} \to \mathbb{R}^{w \times h}$, where d is the number of the parameters of the decoder. In this case, the dynamic of learnable parameters $t \mapsto \omega_\Delta(t)$ should be guided by constraint

$$\Delta(\omega_\Delta(t), A^1(\omega_A(t), t), \ldots, A^n(\omega_A(t), t)) = b(t + \tau) \quad t \in [0, T].$$

In what follows, we will simply write $\Delta(\omega_\Delta(t), A(\omega_A, t))$ in place of the explicit term $\Delta(\omega_\Delta(t), A^1(\omega_A, t), \ldots, A^n(\omega_A, t))$. The incorporation of such constraint is assumed to be done in a soft way using, for example, a quadratic loss $q : \mathbb{R}^{w \times h} \times \mathbb{R}^{w \times h} \to \overline{\mathbb{R}}_+$, with $q(a, b) := (a_{i,j} - b_{i,j})(a_{i,j} - b_{i,j})/2$.

---

[7] Here, we are using the shifted brightness $b(t + \tau)$ instead of the temporal derivative as indicated in Chap. 3 because, while clearly related for suitable choices of τ, this solution seems to be more straightforward to implement.

The other constraint that we dealt with in the previous section, i.e., the brightness invariance, here is replaced by $\bowtie$ relation between the features φ^i and their corresponding velocities v^i. Then, the conjunction relation between the φ^i and the v^i in these new variables can be rewritten as

$$\frac{d}{dt} A^i(\omega_A(t), t) + v^i(\vartheta(t), t) \cdot DA^i(\omega_A(t), t).$$

Please notice here that the total derivative of A^i replaces the partial derivative of $\varphi^i(x, t)$ since A explicitly depends on time both through the input and through the temporal dynamics of the weights. This condition can be enforced by considering the functional $T(\omega_A, \vartheta)$ where [8]

$$(x_1, z) \mapsto T^A(x_1, z) := \int_0^T \left| \frac{d}{dt} A^i(x_1(t), t) + v^i(z(t), t) \cdot DA^i(x_1(t), t) \right|^2 dt.$$

A similar term can be used to impose the constraint on the C field, a possible choice, as it is displayed in Eq. (3.24) would be in that case $T^C(\omega_C, \vartheta)$, where

$$(x_2, z) \mapsto T^C(x_2, z) := \int_0^T \left| \frac{d}{dt} C^j(x_2(t), t) + v^i(z(t), t) \cdot DC^j(x_2(t), t) \right|^2 dt$$

where it is understood that the sum is performed over $i \neq j$. Finally, the B field induces a term $T^C(\omega_C, \vartheta)$ with

$$(x_3, z) \mapsto T^B(x_3, z) := \int_0^T \left| \frac{d}{dt} B^j(x_3(t), t) + (v^i(z(t), t) \right.$$

$$\left. - v^j(z(t), t)) \cdot DB^j(x_3(t), t) \right|^2 dt,$$

where again the summation is performed over $i \neq j$.

The discussion that has been carried on in Chap. 3 shows that beside this defining constraint, in order to have a non-trivial and well defined theory we also need regularization terms. First of all, we need to make sure that the dynamics of the unknowns ω_A, ω_B, ω_C, and ϑ is well behaved; to this end, we introduce the term $S(\omega_A, \omega_B, \omega_C, \omega_\Delta, \vartheta)$ where

---

[8] With the usual conventions for the Einstein summation, and where $| \cdot |$ indicate that we are summing also on the spatial components of A.

$$(x_1, x_2, x_3, y, z) \mapsto \mathcal{S}(x_1, x_2, x_3, y, z) := \frac{1}{2} \int_0^T |x_k''|^2 + |x_k'|^2 + |y'|^2 + |z'|^2.$$

Notice that here we impose an higher regularity on the variables ω_A, ω_B, and ω_C essentially because their derivatives already appear in the terms that enforce the conjugation between A, B, and C and their velocities.

The other natural regularization that we need to impose is the spatial regularization of A, B, and C and to the velocity fields. To this end, we consider the term $\mathcal{W}(\omega_A, \omega_B, \omega_C, \vartheta)$ where $(x_1, x_2, x_3, z) \mapsto \mathcal{W}(x_1, x_2, x_3, z)$ is chosen to be given by

$$\mathcal{W}(x_1, x_2, x_3, z) := \frac{1}{2} \int_0^T \Big(|DA^i(x_1(t), t)|^2 + |DB^i(x_2(t), t)|^2$$

$$+ |DC^j(x_3(t), t)|^2 + |Dv^i(z(t), t)|^2 \Big) \, dt.$$

We are also adding a term $\mathcal{V}(\omega_A, \omega_B, \omega_C)$ that controls the growth of the features by simply weighting the value of A, B, and C:

$$(x_1, x_2, x_3) \mapsto \mathcal{V}(x_1, x_2, x_3)$$

$$:= \frac{1}{2} \int_0^T \Big(|A^i(x_1(t), t)|^2 + |B^i(x_1(t), t)|^2 + |C^j(x_1(t), t)|^2 \Big) \, dt$$

Finally, we need the auto-encoding-like regularization and the terms that propagate this kind of regularization from the A to the B and C field as well as a coercive term on the parameters ϑ. Here, it is convenient for us to group these terms into the functional $\mathcal{I}(\omega_A, \omega_B, \omega_C, \omega_\Delta, \vartheta)$. Here, the functional $(x_1, x_2, x_3, x_4, z) \mapsto \mathcal{I}(x_1, x_2, x_3, x_4, z)$ is defined as follows

$$\mathcal{I}(x_1, x_2, x_3, x_4, z) := \int_0^T \Big(\frac{|z(t)|^2}{2} + \frac{|x_4(t)|^2}{2} + q\big(\Delta(\omega_\Delta(t), A(x_1(t), t)), b(t + \tau)\big)$$

$$+ \Theta(A(x_1(t), t), B(x_2(t), t))$$

$$+ \Upsilon(B(x_2(t), t), C(x_3(t), t)) \Big) \, dt,$$

where the terms $\Theta : \prod_{k=1}^m \mathbb{R}^{w \times h \times n_k} \times \prod_{k=1}^n \mathbb{R}^{w \times h \times m_k} \to \mathbb{R}$ and $\Upsilon : \prod_{k=1}^n \mathbb{R}^{w \times h \times m_k} \times \prod_{k=1}^p \mathbb{R}^{w \times h \times r_k} \to \mathbb{R}$ enforce the regularization presented in Eqs. (3.23) and (3.25), respectively.

With all this definitions laid down, we can now state that the trajectory of the weights $t \mapsto (\omega_A(t), \omega_B(t), \omega_C(t), \omega_\Delta(t), \vartheta(t))$ compatible with a learning process from a visual stream should be constructed so that it optimizes the index

$$\mathcal{S}(\omega_A, \omega_B, \omega_C, \omega_\Delta, \vartheta) + \mathcal{W}(\omega_A, \omega_B, \omega_C) + \mathcal{V}(\omega_A, \omega_B, \omega_C)$$
$$+ \mathcal{I}(\omega_A, \omega_B, \omega_C, \omega_\Delta, \vartheta) + \sum_{\alpha \in \{A, B, C\}} \mathcal{T}^\alpha(\omega_\alpha, \vartheta). \quad (5.9)$$

The previous statement however is vague enough since we have not specified the functional space of trajectories over which we desire to define the optimization problem. As soon as we try to define an appropriate space, we quickly realize that non-causality issues arise from the boundary conditions.

In order to overcome this problem in the next section, we will discuss how to apply a variation of the WIE principle that we used in Sect. 5.2 for the extraction of the optical flow to this problem. This approach as we will see will yield a gradient flow-like dynamics (thus a causal problem). In order to help the intuition on this approach, we will also give a corresponding interpretation in the discrete time.

However, before going on let us rearrange the terms that appear in Eq. (5.9) so that subsequent analysis will be more straightforward. The basic idea is that we want to group together, inside the integral on time, terms that depend on the variables ω_A, ω_B, ω_C, ϑ, and ω_Δ, the terms that depend only on the derivatives of those variables (which is essentially already the $\mathcal{S}$ term), and the terms that contain both the variables and their derivatives (the terms that come from the $\bowtie$ operation). Let $t \mapsto u(t) := (\omega_A(t), \omega_B(t), \omega_C(t), \omega_\Delta(t), \vartheta(t)) \in \mathbb{R}^D$, then the index in Eq. (5.9) is of the form

$$F(u) := \int_0^T \left(T(u'(t), u''(t)) + \frac{1}{2}|Q(u(t))u'(t) + b(u(t))|^2 + V(u(t), t) \right) dt,$$

$$(5.10)$$

where V collects all the terms of $\mathcal{V}$, $\mathcal{W}$, and $\mathcal{I}$, and Q is a (non-square) block matrix and together with b defines the invariance term. Finally T is the rewriting of the term $\mathcal{S}$. It is also useful to rewrite the $\bowtie$ term as follows:

$$\frac{1}{2}|Q(u(t))u'(t) + b(u(t))|^2 = \frac{1}{2}u'(t) \cdot Q(u(t))Q'(u(t))u'(t)$$
$$+ b(u(t)) \cdot Q(u(t))u'(t) + \frac{1}{2}|b(u(t))|^2,$$

which in turn can be rearranged as follows

$$\frac{1}{2}|Q(u(t))u'(t) + b(u(t))|^2 = \frac{1}{2}u'(t) \cdot M(u(t))u'(t) + m(u(t)) \cdot u'(t) + \kappa(u(t)),$$

where this time $\forall x \in \mathbb{R}^D$, $M(x)$ is a square symmetric, positive semi-definite matrix, $m(x) \in \mathbb{R}^D$ and $\kappa(x) \in \mathbb{R}$. This last form is the one that we will use in the next section to write the online learning rules.

5.4 Online Learning

The first step that we need to take in order to obtain causal rules is to appropriately rescale the functional F defined in Eq. (5.10) with the ε factor as we did in Sect. 5.2 for the extraction of the optical flow; in this case, we consider the following family of functionals:

$$
F_\varepsilon(u) := \int_0^T e^{-t/\varepsilon} \left(\frac{\varepsilon\nu}{2} |u'(t)|^2 + \frac{\varepsilon\gamma}{2} u'(t) \cdot M(u(t))u'(t) \quad + \varepsilon\gamma m(u(t)) \cdot u'(t) \right.
$$
$$
\left. + \varepsilon\gamma\kappa(u(t)) + V(u(t), t) \right) dt
$$

where we have chosen $T(a, b) = |a|^2/2$ and we have weighted both the kinetic term and the constraint between the features and the velocities by ε; we have also introduced weights $\nu > 0, \gamma > 0$. Also in this case, the variational problem can be set in the functional space $\mathbb{X}$ defined as in Sect. 5.2 with the only difference that now the co-domain of the elements of $\mathbb{X}$ is $\mathbb{R}^D$ instead of $\mathbb{R}^N$. In order to get local and causal laws, the plan is the same that we discussed for the extraction of the optical flow from the brightness: We write for each fixed ε the Euler–Lagrange equation for F_ε, and then, we let $\varepsilon \to 0$. Using Eqs. (A.5) and (A.6), we obtain the following condition for the minimizer u_ε:

$$
\begin{cases}
- \dfrac{d}{dt} \left(e^{-t\varepsilon} \varepsilon(\nu\mathrm{Id} + \gamma M(u_\varepsilon(t))u'(t) + e^{-t/\varepsilon} \varepsilon\gamma m(u_\varepsilon(t))) \right) \\[2mm]
+ \varepsilon e^{-t/\varepsilon} \left(\dfrac{1}{2} u'_\varepsilon(t) \cdot M'(u_\varepsilon(t))u'_\varepsilon(t) + \gamma m'(u_\varepsilon(t)) \cdot u'_\varepsilon(t) + \kappa'(u_\varepsilon(t)) \right) \quad t \in (0, T) \\[2mm]
+ e^{-t/\varepsilon} \nabla V(u_\varepsilon(t), t) = 0 \\[1mm]
u_\varepsilon(0) = u^0, \quad \varepsilon\big[(\nu\mathrm{Id} + \gamma M(u_\varepsilon(T)))u'_\varepsilon(T) + \gamma m(u_\varepsilon(T)) \big] = 0
\end{cases}
$$
$$(5.11)$$

As $\varepsilon \to 0$, we get the only surviving terms of the Euler equation; i.e., the potential term and those which come from the differentiation of the exponential term are

$$
\begin{cases}
(\nu\mathrm{Id} + \gamma M(u(t)))u'(t) + \gamma m(u(t)) + \nabla V(u(t), t) = 0, \quad t \in (0, T); \\
u(0) = u^0.
\end{cases}
$$

Notice that since $M(x)$ is a positive semi-definite matrix for all $x \in \mathbb{R}^D$, $\nu\mathrm{Id} + \gamma M(x)$ is positive definite and hence invertible. This means that our online update rules can be cast into the form

$$
\begin{cases}
u'(t) = -\big(\nu\mathrm{Id} + \gamma M(u(t))\big)^{-1}\big(\gamma m(u(t)) + \nabla V(u(t), t)\big), \quad t \in (0, T); \\
u(0) = u^0.
\end{cases}
\tag{5.12}
$$

A rather direct interpretation of the above equations can be obtained by considering a discrete time setting and using the idea of minimizing movements (see [2, 39]) with a special similarity term that contains the $\bowtie$ operation. Let (t_k) be a sequence of temporal instants with $t_{k+1} - t_k = \tau > 0$ and consider the following scheme for computing the set of parameters of the model at the next step

$$
\begin{aligned}
u^{k+1} = \arg\min_u \Big(& V_k(u^k) + \nabla V_k(u^k) \cdot (u - u^k) \\
& + \frac{1}{2\tau^2}(u - u^k) \cdot (\nu \mathrm{Id} + \gamma M(u^k))(u - u^k) \\
& + \frac{\gamma}{\tau} m(u^k) \cdot (u - u^k) + \gamma \kappa(u^k) \Big).
\end{aligned}
\tag{5.13}
$$

What we are actually doing here in order to compute the next step is to optimize a second-order approximation of V (here we denote with $V_k(x) := V(x, t_k)$, for all $x \in \mathbb{R}^D$) and a discrete approximation of the $\bowtie$ term. Equation (5.13) implies, just by imposing the stationarity condition of the term inside the $\arg\min$,

$$
u^{k+1} = u^k - \tau^2(\nu \mathrm{Id} + \gamma M(u^k))^{-1}\Big(\frac{\gamma}{\tau} m(u^k) + \nabla V_k(u^k)\Big),
$$

which can indeed be regarded as a discrete approximation of (5.12).

In Sect. 5.2, we showed how a learnable smoothing signal can be introduced to relieve the load of information coming from the environment. Indeed, the very same idea can be used here.

5.5 Online Learning: An Optimal Control Theory Prospective

In Sects. 5.3 and 5.4, we framed the theory described in Chaps. 3 and under the assumption that the model used to compute the feature fields and their corresponding velocities is a "static" FNN whose computational structure is described by Eq. (4.2). At the end of Sect. 4.3, we commented on the fact that we could introduce a more realistic model in which a propagation of visual information along the network is not instantaneous. In what follows, we will only take into account the development of the features φ and their corresponding features and in accordance with the notation introduced in Sect. 5.3 and in Chap. 3 we will denote with $t \mapsto \phi^i(t) \in \mathbb{R}^{w \times h \times m_i}$ the discretized feature trajectories (as usual we will denote with φ the map $t \mapsto (\phi^1(t), \dots, \phi^n(t))$) and with $t \mapsto v^i(t) \in \mathbb{R}^{w \times h \times 2}$ the corresponding *discretized* velocity field. Then, the computational model described in Eq. (4.5) could be more generally rewritten as

$$
\dot{\varphi}(t) = f(\varphi(t), u(t), a(t), t)
\tag{5.14}
$$

where, consistently with the notation introduced in Sect. 5.3, $u(t)$ is the set of parameters of the network and $a(t)$ is the focus of attention, which is a crucial information to have in order to select the parameters in u for the FNN computation. The explicit dependence on t takes into account the dependence on the visual signal (i.e., the brightness b). The map f is related to the function F of Eq. (4.5) but of course acts on the newly arranged and discretized variables.

Correspondingly for the velocities, we could write a similar model:

$$\dot{\mathbf{v}}(t) = g(\mathbf{v}(t), u(t), a(t), t). \tag{5.15}$$

Here, the function g is the counterpart of the function f in Eq. (5.14) and it aggregates the information at a certain level of the network that estimate the conjugate velocities to compute the field at a higher level.

Equations (5.14) and (5.15) basically tell you how to compute the features and the corresponding velocities once the parameters u are given, and the learning process is indeed the process of optimal selection of these parameters. The meaning of "optimal" as we saw in Sect. 5.3 can be specified in terms of the value of *running cost* r at time t which depends of course on the value of φ, $\mathbf{v}$, and u and that therefore can be thought as a function $r : \mathbb{R}^{w \times h \times \sum_i m_i} \times \mathbb{R}^{w \times h \times 2 \times n} \times \mathbb{R}^D \times [0, T] \to \mathbb{R}$ which specify the cost $r(\varphi(t), \mathbf{v}(t), u(t), t)$, through which we would like to formulate the learning problem as the minimum problem:

$$\inf_{u \in X} \int_0^T r(\varphi(t), \mathbf{v}(t), u(t), t) \, dt.$$

Such cost is the rewriting in this new variables of the potential term V plus the invariance term on the features. It is worth noticing that, as we already remarked in Chap. 3, the invariance term $\dot{\phi}^i + v^i \cdot D\phi^i = 0$ in virtue of Eq. (5.14) could be enforced by a penalty term

$$\sum_{i=1}^n (f^i(\varphi, u(t), t) + v^i(t) \cdot D\phi^i(t))^2.$$

Stated in this terms the learning problem can indeed be regarded as an optimal control problem (see [29]) in which

- φ and $\mathbf{v}$ represent what is usually called the *state* of the system.
- The learnable parameters u are the control parameters that should be chosen compatibly with the state dynamic and with the cost r.

It is clear then that this point of view opens new and broad prospective to the study of the online learning process. Among them, a particularly promising direction is that of addressing the optimal-control-like problem described above using the methods of *dynamic programming* that lead to the study of Hamilton–Jacobi equations.

Finally, we believe that is important to notice that, in fact, also the focus of attention could in principle be considered a sort of "control parameter" and could be incorporated in the vectorial variable u. In this case, as it is described in Sect. 2.5 the equations that determine the trajectory $t \mapsto a(t)$ will be coupled with those that perform feature extraction.

5.6 Why is Baby Vision Blurred?

There are surprising results from developmental psychology on what newborns see. Basically, their visual acuity grows gradually in early months of life. Interestingly, Charles Darwin had already noticed this very interesting phenomenon. In his own words:

> It was surprising how slowly he acquired the power of following with his eyes an object if swinging at all rapidly; for he could not do this well when seven and a half months old.

At the end of the seventies, this early remark was given a technically sound basis (see, e.g., [27]). In that paper, three techniques—optokinetic nystagmus (OKN), preferential looking (PL), and the visually evoked potential (VEP)—were used to assess visual acuity in infants between birth and six months of age. More recently, Braddick and Atkinson [21] provide an in-depth discussion on the state of the art in the field. It is clearly stated that for newborns to gain adult visual acuity, depending on the specific visual test, several months are required. Then, we are back to **Q9** of Sect. 1.6 on whether such an important cognitive process is a biological issue or if it comes from higher information-based laws of vision. Overall, the blurring process pops up the discussion on the protection of the learning agent from information overloading, but we want to deepen this topic at the light of the vision field theory of this book.[9]

While the enforcement of vision field theory constraints onto foveated neural networks results into a clean variational problem, its numerical solution might be hard. The solution herein presented in Sect. 5.4 very well resembles classic machine learning approaches based on gradient descent. However, it is worth mentioning that we are in front of an online gradient computation that should not be confused with stochastic gradient, for which there is a lot of experimental evidence on its behavior as well as a number of significant theoretical statements on its convergence. Unfortunately, when working online the specter of forgetting behavior is always lurking! The learning agent might adapt his weights significantly when looking at a still image for days, thus forgetting what he learned earlier. The formulation given in Sect. 5.5 considers the actual minimization of the functional but, as yet, its effective solution is an open problem. Hence, the "solution" that biology has discovered for children deserves investigation.[10]

In Sect. 5.2, we have introduced the basic idea behind video blurring in the case of prediction of the optical flow. While this is a dramatically simplified version of

---

[9] Online learning on video is mostly unexplored!

[10] Blurring the video for optical flow and beyond.

the overall theory, the problem already presents the seed for the formulation of an appropriate solution in the general case. The input undergoes a smoothing process, the purpose of which is to "simplify life" at the beginning of the agent's life. In general, if we remove the information from the source by massive low-pass filtering the problem of learning has a trivial solution. The essence of the idea is that of beginning from such a solution and then learns by tracking the constraints by properly controlling the smoothing parameter (σ in Sect. 4.4) that is expected to converge to zero at the end of learning. In general, one can rely on additional variables to prevent from information overloading with the purpose of stabilizing the dynamical system, which tracks the injection of information. The development of any computation model that adheres to this view is based on modifying the connections along with an appropriate input filtering so as the learning agent always operates at an equilibrium point [10].[11]

When looking at the foveated networks, particularly to filter g, one can promptly recognize a natural access for smoothing the signal, since we can play with the control variable σ of Eq. (4.3). The role of blurring is likely connected with the degree of quality of the expected visual skills. The higher the quality, the higher such a process is likely required. The effect of smoothing clearly simplifies stability issues of the learning equations that are more serious when the vision field graph become more and more complex. While as already pointed out, chicks in the first days of life find it more difficult to see under quick movements, children do require much more time for conquering adult's visual skills. At the birth, chicks and children rely on the genetic transmission of information for their visual skills. When considering space constraints, children likely use a lot of that space for coding the learning mechanism with respect to chicks.[12]

In Sect. 3.7, we investigated the nature of vision fields and their relationship with classic field theory with specific reference to electromagnetism. While the mutual refinement of the $E - B$ fields holds at any frequency, as we set a threshold for neglecting signals below its choice, we clearly see that the number of refinements (number of terms of the Bessel's function expansion) is strongly dependent on the chosen frequency. Notice that, because of the superposition principle, the signal can also be carried out by supplying an input with rich spectral structure to the capacitor. Such a signal contains information that is related to such a composite structure, and it has an interesting analogy with the video signal. Now, we notice that one could elaborate the described Feynman's analysis by assuming that the signal at the capacitor begins from DC and gradually moves to its final spectral structure. The development of Bessel's term from the $E - B$ interaction is dual to the development of the vision field interaction, which suggests that the gradual exposition to the information might simplify the learning process.

---

[11] Blurring and foveated nets.

[12] Feynman's discovery of Bessel's equation in capacitors and blurring processes.

When promoting the role of time, the arising prealgorithmic framework suggests extending the learning process to an appropriate modification of the input
that is finalized to achieve the expected "visual acuity" at the end of the process
of learning. This likely stabilizes the learning process especially in complex
foveated animals. Hence, as visual tasks involve the presence of highly structured objects—like for humans—the necessity of FOA leads to more complex
learning models that likely benefit from video blurring.

Chapter 6
Non-visual Environmental Interactions

Under normal conditions, the research scientist is not an innovator but a solver of puzzles, and the puzzles upon which he concentrates are just those which he believes can be both stated and solved within the existing scientific tradition.
Thomas Kuhn, "The Structure of Scientific Revolution", 1962

Keywords Object recognition · Language · Learning en plein air

6.1 Object Recognition and Related Visual Skills

This books has covered information extraction from visual sources as a consequence of the vision field theory. Animals and machines living in a certain environment, however, are expected to carry out specific tasks that might involve actions and/or perceptions. In both cases, visual agents receive precious information which drives the learning process accordingly. The supervised learning protocol that has been dominating computer vision for classification offers massive labeling of the training set, so as the non-visual environmental interaction results in the systematic labeling of each single example. All animals, including humans, experiment truly different environmental interactions, which typically convey much less information with respect to the supervised learning protocol. The predator–prey ordinary interactions are not restricted to visual signals, and, likewise, their movement produces precious feedback for appropriate navigation plans. When shifting attention to humans, things become more sophisticated and involve specific perceptual issues that seem to characterize our more advanced abstraction capabilities.

Object recognition

How can we recognize objects without massive supervision? How can children associate the name to an object after a few explicit supervisions based on its pointing? The surprising limitations in frogs' visual skills suggest to better analyze the fundamental role of eye movements and focus of attention, which is mostly missing in frogs. Likewise, there is nothing similar to object supervision based on pointing to

© The Author(s), under exclusive license to Springer Nature Switzerland AG 2022
A. Betti et al., *Deep Learning to See*, SpringerBriefs in Computer Science,
https://doi.org/10.1007/978-3-030-90987-1_6

a specific pixel in the retina, which clearly requires FOA. As discussed in this book, FOA is interwound with the development of visual features with different degree of abstraction until the interpretation of objects is gained. As such, it necessarily plays a crucial role for object recognition, which does not typically require the full discovery of all objects at a given frame. On the opposite, humans experiment a cognitive process that is driven by the FOA, which leads to explore all the visual scene which is relevant from an information viewpoint. Of course, just like classic convolutional networks, the computational model driven by the focus of attention, referred to as foveated networks in this book, can compute features and object categories on any pixel of the retina.[1]

In order to discuss object recognition, we begin facing the following apparently trivial question: What is an object? Of course, it is connected with its visual interpretation. When trying to answer this question, we early realize that we need to unfold the semantics of objects. While they can be described in terms of visual features, objects are in fact characterized by their physical structure, which is the distinguishing property that has not been considered so far for visual features. The physical structure manifests itself through the connectedness of the composing atoms, which leads to regard objects as single entities. The definition of the features emerging from the two principles of visual perception does not reflect such a property that objects possess. While interaction features ψ and χ gain properties connected with the object affordance, it is quite obvious that they can only represent parts of an objects, since they are learned without transmitting the above-mentioned object physical connectedness property. Objects are like puzzles composed of different pieces (features); such a structured composition generates visual patterns whose composition needs to be disclosed to the visual learning agent.

As already mentioned in Sect. 3.3, the attachment to the earth, which leads to the distinction between attached and detached objects, results in a remarkable difference in their affordance and in the requirement of environmental interactions for gaining recognition.[2] Detached objects can move with respect to the background, which provides information for capturing directly the above-mentioned property. The movement of only one object with respect to the background can be detected easily by subtracting the velocity of the FOA. The estimation of the kinematic velocity makes it possible to localize the pixels where the object covers the retina. Some complex movements, which result in nearly null velocity of points of the object, do not compromise the acquisition of the object. Basically, in the special case of a single object moving with respect to the background, there is an explicit information coming from motion which characterizes the object. The optical flow in fact selects the pixels of the moving object, so as to get semantic labeling for free.

[3] The recognition of attached object presents a higher degree of complexity. We also need to bear in mind that while some objects are attached in strict sense (e.g. think of houses of a village), even though others are not, they are visually similar to attached

---

[1] Objects' distinguishing properties and their recognition.

[2] Detached objects.

[3] Attached objects.

objects in their visual environments since they rarely move. Unlike detached objects there is no optical flow which transmits evidence on their identity. When looking at a pine tree from tens of meters, one perceive it thanks to its prominent trunk and to the green needles. There is no motion information which can aggregate trunk and pine needles, so as we can perceive it as unique connected entity. The trunk could in fact be aggregated with the background without providing any additional cognitive evidence. Overall, complex objects which manifest themselves by many different parts seem to be more difficult to be recognized without involving explicit information on their structure. As already mentioned, the FOA is the crucial mechanism which drives the recognition, especially in this case.

A visual agent can ask questions on which object(s) is (are) located in the pixel where he focuses attention, which results in a very precious supervision. Of course, also the supervisor can take the initiative of providing the supervision, but this works only provided that the agent has already reached a certain degree of learning of motion invariant features (Fig. 6.1). A fundamental results that emerges from the theory on

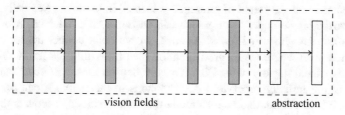

vision fields   abstraction

Fig. 6.1 Weights of the abstract module are learned in the second developmental stage after having frozen the weights of the neurons that implement the vision fields

vision fields is that, as we provide the supervision on the FOA, the learning process involves the weights a used to properly select the visual features that concur to the recognition. Basically, we have

$$o_i = \sigma \left(\sum_j a_{ij} \chi_j \right), \tag{6.1}$$

where we learn the weights a_{ij} with the specific supervision on the FOA. This equation clearly unfolds the fundamental difference in the training process that involves the extraction of high-level cognitive concepts, like structured attached objects, with respect to the visual features χ_j.

In the case in which χ_j is a feature group that is expected to be conjugated with the velocity field v, also o_i is conjugated with the same field, which can clearly explain the segmentation capability. The localization of the pixels of the object that are typically based on thresholding criteria on a_i are in fact corresponding with a related computation on the associated velocity field v that allows us to identify the pixels covered in the retina by the object. Interestingly, the theory guarantees the consistency with what can be extracted by o_i.

Of course, we can also construct computational units for detecting objects o_i which are based on feedforward-based computations described by $\vdash o_i$ and, therefore, we can consider o_i as a vision field that comes from different feature groups. This makes sense, especially when considering highly structured objects. The corresponding conjugation with different velocity fields suggests that the overall process of object recognition, as perceived by humans, needs to fully exploit the traditional supervision that is experimented with children. We simply provide supervision on the presence of objects by pointing them straight or, alternatively, we answer children's questions on objects name. As discussed later on, the communication protocol is far richer than a simple interaction on the specific point of FOA, but facing this case is very useful since it also helps addressing more general linguistic interactions.[4]

Hence, let us analyze the fundamental role of FOA that arises when we want to recognize objects. The single supervision of an object in the FOA results in fact in a collection of related supervisions that are gained as the FOA moves in the retina. Let us assume that the frame of reference is located in the FOA. Basically, we can memorize the single supervision which can be transferred, later on, to all pixels related to the movement of focus. As such, we virtually collect frames referring to the same pixel where we focused attention seen from different frame references generated by the trajectory of the FOA. This mechanism for supervision transferring holds until the initially supervised pixel remains in the retina. It can even move itself, but it is important that its location in the retina in order to activate the object recognition on its position from the current FOA.

This analysis strongly supports the need of FOA for achieving top-level object recognition performance. Whenever FOA is missing, like in frogs, we are missing the possibility of acquiring a lot of supervision for free. This holds also for convolutional neural networks, whose remarkable results seem to be due to the exploitation of the supervised learning protocol that is made possible in machines. Outside that battlefield, when the environmental interactions are restricted to those typically experimented in nature, computational models of learning based on convolutional nets might be quite limited.

6.2 What Is the Interplay with Language?

The interplay of vision and language is definitely one of the most challenging issues for an in-depth understanding of human vision. While the vision field theory presented in this book can be used as a basis for understanding from a functional viewpoint the visual skills in most animal species, humans exhibit a sophisticated behavior in the interplay between vision and language that is only superficially covered here.

---

[4] FOA-based supervision transferring: A single supervision on the FOA of attached objects results in a collection of supervisions.

On the other hand, along with the associated successes, the indisputable adoption of the supervised learning protocol in most challenging object recognition problems caused the losing of motivations for an in-depth understanding of the way linguistic information is synchronized with visual clues. In particular, the way humans learn the name of objects is far away from the current formal supervised protocol. This can likely be better grasped when we begin considering that top-level visual skills can be found in many animals (e.g., birds and primates), which clearly indicates that their acquisition is independent of language.

Hence, as we clarify the interplay of vision and language we will likely address also the first question on how to overcome the need for "intensive artificial supervision." Since first linguistic skills arise in children when their visual acuity is already very well developed, there is a good chance that simple early associations between objects and their names can easily be obtained by "a few supervisions" because of the very rich internal representation that has already been gained of those objects. It is in fact only a truly independent hidden representation of objects which makes their subsequent association with a label possible! This seems to be independent of biology, whereas it looks like a fundamental information-based principle, which somehow drives the development of "what" neurons.

The interplay of language and vision has been recently very well addressed in a survey by [69]). It is claimed that performance on tasks that have been presumed to be non-verbal is rapidly modulated by language, thus rejecting the distinction between verbal and non-verbal representations. While we subscribe the importance of sophisticated interactions, we also reinforce the claim that identifying single objects is mostly a visual issue. However, when we move toward the acquisition of abstract notions of objects, the interaction with language becomes more and more important.

Once again, the discussion carried out so far promotes the idea that for a visual agent to efficiently conquer the capabilities of recognizing objects from a few super-visions, it must undergo some developmental steps aimed at developing invariant representations of objects, so as the actual linguistic supervision takes place only after the development of those representations. But, when should we enable a visual agent to begin with the linguistic interaction? While one might address this question when attacking the specific computational model under investigation, a more natural and interesting way to face this problem is to re-formulate the question as:

Q10: *How can we develop "linguistic focusing mechanisms" that can drive the process of object recognition?*

This is done in a spectacular way in nature! Like vision, language development requires a lot of time. Interestingly, it looks like it requires more time than vision. The discussion in Sect. 5.6 indicates that the gradual growth of the visual acuity is a possible clue to begin with language synchronization. The discussed filtering process offers a protection from visual information overloading that likely holds for language as well. As the visual acuity gradually increases, one immediately realizes that the mentioned visual language synchronization has a spatiotemporal structure. At a certain time, we need to inform the agent on what we see at a certain position in the retina. As already mentioned, an explicit implementation of such an association

Fig. 6.2 Developmental learning to see. First vision fields are developed according what is stated in the I and II principles by following the arrows. The object acquisition involves the additional pointing info on the FOA, while a virtuous loop with scene interpretation also reinforces object recognition

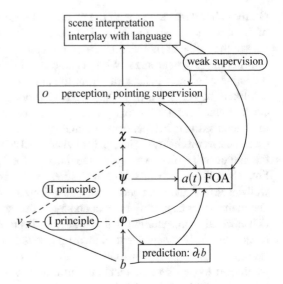

can be favored by an active learning process: The agent can ask itself what is located at (x, t). However, what if you cannot rely on such a precious active interactions? For example, a linguistic description of the visual environment is generally very sophisticated and mentions objects located in different positions of the retina, without providing specific spatiotemporal information.

Basically, this is a sort of *weak supervision* that is more difficult to grasp. However, once again, developmental learning schemes can significantly help. At early stage of learning, the agent's tasks can be facilitated by providing spatiotemporal information. For example, naming the object located where the agent is currently focusing attention conveys information by a sort of human-like communication protocol. As time goes by, the agent gains gradually the capability of recognizing a few objects. What really matters is the confidence that is gained in such a task. When such a developmental stage is reached, linguistic descriptions and any sort of natural language-based visual communication can be conveniently used to reinforce the agent recognition confidence. Basically, these weak supervisions turn out to be very useful since they can profitably be attached where the agent came up with a prediction that matches the supervision. The acquisition of the capability of recognizing objects described in the previous section can in fact fire a subsequent reinforcement. If the agent detects the presence of a certain word on a given frame and he can detect an object with that name, he can add such an information as an additional supervision, along with the correspondent compatible frames in which the object is still present in the retina.[5]

In the above discussion, the weak supervision, which takes place without an explicit "pointing information," is used for reinforcing concepts that have been

---

[5] Supervised learning from weak supervision.

acquired by supervision on the FOA. Can weak supervision also fire a learning process without leaning on pointing information? This is far more complex. The only possibility for the agent is that of discovering abstract reduced features χ that have not been labelled yet and attach them the candidate supervision. In principle, this is possible cognitive path that, however, is remarkably more difficult to follow with respect to learning the name of an object in the FOA.[6]

Overall, the investigation path followed in this book is disseminated of a number of insights which seem to provide evidence that stage-based learning to see in animals is mostly an information-based issue which holds regardless of biology. Figure 6.2 reports a summary of the drawn analysis which provide some evidence on the claimed role of time and on the need to undergo developmental stages. The stage transition is especially evident as we involve the learning agent into the task of object recognition which requires additional information with respect to the video source. In particular, the weight of learning is independent of what has been obtained in the previous stages on the construction of visually features invariant under motion.[7]

Computer vision and natural language processing have been mostly evolving independently one each other. While this makes sense, the time has come to explore the interplay between vision and language with the main purpose of going beyond the protocol of supervised learning for attaching labels to objects. Interestingly, challenges arises in scene interpretation when we begin considering the developmental stages of vision that suggest gaining strong object invariance before the attachment of linguistic labels.

6.3 The "en Plein Air" Perspective

Posing the right questions is the first fundamental step to gain knowledge and solve problems. Hopefully, the questions raised in this book might give the reader insights and contribute to face fundamental problems in computer vision. However, one might wonder what could be the most concrete action for promoting studies on the posed questions. So far, computer vision has strongly benefited from the massive diffusion of benchmarks which, by and large, are regarded as fundamental tools for performance evaluation. However, it is clear that they are very well-suited to support the statistical machine learning approach based on huge collections of labelled images. However, this book opens the doors to explore a different framework for performance evaluation. The emphasis on video instead of images does not lead us to think of huge collection of video, but to a truly different approach in which no collection at all

---

[6] Can it learn object names without explicit "pointing supervision"?.

[7] Developmental learning to see.

is accumulated! Just like humans, machines are expected to live in their own visual environment. However, what is the scientific framework for evaluating the performance and understand when a theory carries out important new results? Benchmark bears some resemblance to the influential testing movement in psychology which has its roots in the turn-of-the-century work of Alfred Binet on IQ tests ([16]). Both cases consist in attempts to provide a rigorous way of assessing the performance or the aptitude of a (biological or artificial) system, by agreeing on a set of standardized tests which, from that moment onward, became the ultimate criterion for validity. On the other hand, it is clear that the skills of any visual agent can be quickly evaluated and promptly judged by humans, simply by observing its behavior. How much does it take to realize that we are in front of person with visual deficits? Do we really need to accumulate tons of supervised images for assessing the quality of a visual agent? The clever idea behind ImageNet [26] is based on crowdsourcing. Could not we also use crowdsourcing as a *crowdsourcing performance evaluation scheme*? People who evaluate the performance could be properly registered so as to limit spam (see e.g. [43]).

Scientists in computer vision could start following a sort of term *en plein air*, the term which is used to mimic the French Impressionist painters of the nineteenth century and, more generally, the act of painting outdoors. This term suggests that visual agents should be evaluated by allowing people to see them in action, virtually opening the doors of research labs.

While the idea of shifting computer vision challenges into the wild deserve attention, one cannot neglect the difficulties that arise from the lack of a true lab-like environment for supporting the experiments. The impressive progress in computer graphics, however, is offering a very attractive alternative that can dramatically facilitate the developments of approaches to computer vision that are based on the online treatment of the video (see e.g. [74]).

Needless to say, computer vision has been fueled by the availability of huge labeled image collections, which clearly shows the fundamental role played by pioneering projects in this direction (see, e.g., [26]). The ten questions posed in this paper will likely be better addressed only when scientists will put more emphasis on the en plein air environment. In the meantime, the major claim of this book is that the experimental setting needs to move to virtual visual environments. Their photorealistic level along with explosion of the generative capabilities makes these environment just perfect for a truly new performance evaluation of computer vision. The advocated crowdsourcing approach might really change the way we measure the progress of the discipline.

Appendix A
Calculus of Variations

Calculus of variation is in its essence the study of extremals of functions $f : \mathbb{X} \to \overline{\mathbb{R}}$, where $\overline{\mathbb{R}} = \mathbb{R} \cup \{-\infty, +\infty\}$. The case in which $\mathbb{X}$ is a Euclidean space corresponds of course to the study of stationary points of a real-valued function on $\mathbb{R}^n$. For the purposes of this work, we are mainly interested in the case in which $\mathbb{X}$ is an infinite-dimensional functional space. In particular, we will focus on the case in which $\mathbb{X}$ is an affine space with vector space V so that in particular for all $x \in \mathbb{X}$ and all $v \in V$ we have that $x + v \in \mathbb{X}$. In this case, then it is particularly straightforward to generalize the usual concept of directional derivative in the direction v at point x_0 as follows:

$$\delta F(x_0, v) := \lim_{s \to 0} \frac{F(x_0 + sv) - F(x_0)}{s}. \tag{A.1}$$

In general, this quantity is called *Gâteaux differential* or more traditionally *first variation*. The term Gâteaux differential comes from the notion of Gâteaux differentiability in Banach spaces (see [36]). Notice also that if we define $\psi(s) := F(x_0 + sv)$, then $\delta F(x_0, v) = \psi'(0)$.

This quantity is particularly important for the study of extremals of a functional since, as it happens for real-valued functions, the vanishing of this quantity for all $v \in V$ it is a necessary condition to be satisfied by any local extremum of F.

A.1 Integral Functional and Euler Equations

We will now restrict ourselves to functionals of the form

$$F(x) := \int_0^T L(t, x(t), \dot{x}(t)) \, dt, \tag{A.2}$$

A. Betti et al., *Deep Learning to See*, SpringerBriefs in Computer Science,
https://doi.org/10.1007/978-3-030-90987-1_A

where $x \in \mathbb{X}$ and $L(t, z, p)$ is a continuous real-valued function of the variables (t, z, p). Suppose furthermore that, for example,[1] $\mathbb{X} \subset C^2([0, T]; \mathbb{R}^n)$.

Now we can use the following well-known result about integration that essentially says that under appropriate regularity assumptions *the derivative of the integral is the integral of the derivative*. More precisely if $f : [0, T] \times [-\tau, \tau] \to \mathbb{R}$, then if we let $\psi(s) := \int_0^T f(t, s) \, dt$ we have that

1. If f is continuous in $[0, T] \times [-\tau, \tau]$, then ψ is continuous in $[-\tau, \tau]$.
2. If f_s is continuous in $[0, T] \times [-\tau, \tau]$, then ψ' exists and it is given by

$$\psi'(s) := \int_0^T f_s(x, s) \, dx.$$

If we take, as we remarked above $\psi(s) := F(x + sv)$, then

$$\psi'(s) = \int_0^T \frac{d}{ds} L(t, x + sv, \dot{x} + s\dot{v})$$

Then, using the chain rule on L we have that the first variation (A.1) looks like

$$\delta F(x, v) = \int_0^T \left(L_z(t, x(t), \dot{x}(t)) \cdot v(t) + L_p(t, x(t), \dot{x}(t)) \cdot \dot{v}(t) \right)$$

Now, consider the relation $\delta F(x, v) = 0$ for all $v \in C^\infty$; here, we are considering $v \in C^\infty$ instead of C^2 functions since this has the advantage that we can consider the same class of variations for all differential equations of all order; moreover, it is consistent with the usual conventions in the theory of distributions. This condition is equivalent to

$$\int_0^T \left(L_z(t, x(t), \dot{x}(t)) \cdot v(t) + L_p(t, x(t), \dot{x}(t)) \cdot \dot{v}(t) \right) dt = 0 \quad \forall v \in C^\infty((0, T); \mathbb{R}^n)$$

This condition is usually called the weak Euler equation for x. Notice that in order for this condition to be well defined the function $x(t)$ does not need more regularity than that declared in the definition of $\mathbb{X}$.

Now suppose that we take v vanishing at the boundary, then by integration by parts we get

---

[1] A milder assumption would be $\mathbb{X} \subset H^2((0, T); \mathbb{R}^n)$.

$$\int_0^T \left(L_z(t, x(t), \dot{x}(t)) - \frac{d}{dt} L_p(t, x(t), \dot{x}(t)) \right) \cdot v(t) \, dt = 0. \tag{A.3}$$

This, of course, can be done if we have enough regularity on both L and x. This integral relation can be turned into a differential equation by using the fundamental lemma of calculus of variations:

Lemma A.1 *Let* $f : [0, T] \to \mathbb{R}^n$ *be continuous function. If*

$$\int_0^T f(t) \cdot v(t) \, dt = 0 \quad \forall v \in C_c^\infty((0, T); \mathbb{R}^n), \tag{A.4}$$

then $f \equiv 0$ *in* $[0, T]$.

We therefore have

$$L_z(t, x(t), \dot{x}(t), \ddot{x}(t)) - \frac{d}{dt} L_p(t, x(t), \dot{x}(t), \ddot{x}(t)) = 0. \tag{A.5}$$

Till now, we have not taken into account the cases in which the variations do not vanish at the boundary; this, for example, happens when if we formulate a variational problem without specifying in $\mathbb{X}$ the value of the solution at the boundaries.

Notice however that since the fundamental lemma of the calculus of variation the vanishing condition (A.4) needs just to be verified for compactly supported functions; if we can prove that the solution of the variational problem is regular enough, then the differential Eq. (A.5) will hold regardless of any boundary conditions. This being said let us now see what happens to the stationarity condition $\delta F(x, v) = 0$ when we do not assume the vanishing of the variation at the boundaries.

We have already discussed the fact that Eq. (A.5) still holds; therefore, we are left with the contributions only from the boundary terms of the integration by parts; namely,

$$\left[L_p \cdot v \right]_0^T = 0.$$

In order for this term to be zero, the only possibility other than the vanishing of v is to have

$$L_p = 0, \tag{A.6}$$

at the boundary where we do not know that the variation is vanishing. This kind of boundary conditions usually is referred to as Neumann boundary conditions since they generally depend on the derivatives of the solution.

References

1. Allport, A.: Visual attention. In: Foundations of Cognitive Science, pp. 631–682. The MIT Press (1989)
2. Ambrosio, L., Maso, G.D., Marco, F., Miranda, M., Spagnolo. S.: Ennio De Giorgi-Selected Papers. Springer, Berlin (2006)
3. Baker, N., Erlikhman, G., Kellman, P.J., Lu, H.: Deep convolutional networks do not perceive illusory contours. In: Kalish, C., Rau, M.A., (Jerry) Zhu, X., Rogers, T.T. (Eds.), Proceedings of the 40th Annual Meeting of the Cognitive Science Society, CogSci 2018, Madison, WI, USA, July 25–28, 2018. cognitivesciencesociety.org (2018)
4. Baker, S., Scharstein, D., Lewis, J.P., Roth, S., Black, M.J., Szeliski, R.: A database and evaluation methodology for optical flow. Int. J. Comput. Vis. **92**(1), 1–31 (2011)
5. Dana, H.: Ballard. Animate vision. Artif. Intell. **48**(1), 57–86 (1991)
6. Beltran, W.A., Cideciyan, A.V., Guziewicz, K.E., Iwabe, S., Swider, M., Scott, E.M., Savina, S.V., Ruthel, G., Stefano, F., Zhang, L., Zorger, R., Sumaroka, A., Jacobson, S.G., Aguirre, G.D.: Canine retina has a primate fovea-like bouquet of cone photoreceptors which is affected by inherited macular degenerations. PLOS ONE **9**, 1–10
7. Tatler Benjamin, W., Hayhoe Mary, M., Land Michael, F., Ballard Dana, H.: Eye guidance in natural vision: reinterpreting salience. J. Vis. **11**, 1–23 (2011)
8. Bertasius, G., Wang, H., Torresani, L.: The directional tuning of the barber-pole illusion. arXiv:2102.05095, 2021
9. Betti, A., Gori, M.: Convolutional networks in visual environments (2018). Arxiv preprint arXiv:180107110v1
10. Betti, A., Gori, M., Melacci, S.: Learning and visual blurring. Technical Report. SAILab (2021)
11. Betti, A.: A variational framework for laws of learning. Ph.D. thesis, vol. 2 (2020). http://hdl.handle.net/2158/1185278
12. Betti, A., Gori, M.: The principle of least cognitive action. Theor. Comput. Sci. **633**, 83–99 (2016)
13. Betti, A., Gori, M.: Backprop diffusion is biologically plausible. CoRR (2019). abs/1912.04635
14. Betti, A., Gori, M.: Backprop diffusion is biologically plausible (2020)
15. Betti, A., Gori, M., Melacci, S.: Cognitive action laws: the case of visual features. CoRR (2018). abs/1808.09162
16. Binet, A., Simon, T.: The Development of Intelligence in Children: The Binet? Simon Scale. Williams & Wilkins (1916)
17. Boccignone, G., Cuculo, V., D?Amelio, A.: Problems with saliency maps. In: International Conference on Image Analysis and Processing, pp. 35–46. Springer, Berlin (2019)

© The Author(s), under exclusive license to Springer Nature Switzerland AG 2022
A. Betti et al., *Deep Learning to See*, SpringerBriefs in Computer Science,
https://doi.org/10.1007/978-3-030-90987-1

18. Boccignone, G., Ferraro, M.: Modelling gaze shift as a constrained random walk. Physica A **331**(1–2), 207–218 (2004)
19. Borenstein, E., Ullman, S.: Class-specific, top-down segmentation. In: Heyden, A., Sparr, G., Nielsen, M., Johansen, P. (eds.) Proceedings of Computer Vision—ECCV 2002, 7th European Conference on Computer Vision, Copenhagen, Denmark, May 28–31, 2002, Part II, vol 2351, Lecture Notes in Computer Science, pp. 109–124. Springer, Berlin (2002)
20. Borji, A., Itti, L.: State-of-the-art in visual attention modeling. IEEE Trans. Pattern Anal. Mach. Intell. **35**(1), 185–207 (2012)
21. Braddick, O., Atkinson, J.: Development of human visual function. Vis. Res. **51**, 1588–1609 (2011)
22. Briggs, F., Martin Usrey, W.: A fast, reciprocal pathway between the lateral geniculate nucleus and visual cortex in the macaque monkey. J. Neurosci. **27**(20), 5431–5436 (2007)
23. Buiatti, M., Di Giorgio, E., Piazza, M., Polloni, C., Menna, G., Taddei, F., Baldo, E., Vallortigara, G.: Cortical route for facelike pattern processing in human newborns. PNAS (2019)
24. Chen, S., Zhao, Q.: Boosted attention: leveraging human attention for image captioning. In: Proceedings of the European Conference on Computer Vision (ECCV), pp. 66–84 (2018)
25. Crick, F.: The recent excitement about neural networks. Nature **337**, 129–132 (1989)
26. Deng, J., Dong, W., Socher, R., Li, L.J., Li, K., Fei-Fei, L.: ImageNet: a large-scale hierarchical image database. In: CVPR09 (2009)
27. Dobson, V., Teller, D.Y.: Visual acuity in human infants: a review and comparison of behavioral and electrophysiological studies. Vis. Res. **18** (1978)
28. Dyson, F.J.: Infinite in All Directions: Gifford Lectures Given at Aberdeen, Scotland April–November 1985. HarperCollins (2004)
29. Evans, L.C.: Partial differential equations. Am. Math. Soc. **19** (2010)
30. Faggi, L., Betti, A., Zanca, D., Melacci, S., Gori, M.: Wave propagation of visual stimuli in focus of attention. CoRR. abs/2006.11035 (2020)
31. Farroni, T., Johnson, M.H., Menon, E., Zulian, L., Faraguna, D., Csibra, Gergely: Newborns' preference for face-relevant stimuli: effects of contrast polarity. Proc. Natil. Acad. Sci. **102**(47), 17245–17250 (2005)
32. Feynman, R.P., Leighton, R.B., Sands, M..: The Feynman Lectures on Physics, Vol. II: The New Millennium Edition: Mainly Electromagnetism and Matter. The Feynman Lectures on Physics. Basic Books (2011)
33. Fisher, N., Zanker, J.M.: Is space-time attention all you need for video understanding? Perception 1321–1336 (2001)
34. Gall, J., Fossati, A., Van Gool, L.: Functional categorization of objects using real-time markerless motion capture. In: The 24th IEEE Conference on Computer Vision and Pattern Recognition, CVPR 2011, Colorado Springs, CO, USA, pp. 1969–1976. IEEE Computer Society (2011)
35. Gerkema, M.P., Davies, W.I.L., Foster, R.G., Menaker, M., Hut, R.A.: The nocturnal bottleneck and the evolution of activity patterns in mammals. Proc. R. Soc. Lond. Ser. B Biol. Sci. **280**(1765) (2013)
36. Giaquinta, M., Hildebrandt, S.: Calculus of Variations I. Grundlehren der mathematischen Wissenschaften. Springer, Berlin Heidelberg (2004)
37. Gibson, J.J.: The Senses Considered as Perceptual Systems. Houghton Mifflin, Boston (1966)
38. Gibson, J.J.: The Ecological Approach to Visual Perception. Houghton Mifflin, Boston, Boston (1979)
39. Gobbino, M.: Minimizing movements and evolution problems in Euclidean spaces. Annali di Matematica pura ed applicata **176**(1), 29–48 (1999)
40. Goodale, M.A., Keith Humphrey, G.: The objects of action and perception. Cognition **67**(1), 181–207 (1998)
41. Goodale, M.A., Milner, A.D.: Separate visual pathways for perception and action. Trends Neurosci. **15**(1), 20–25 (1992)
42. Gori, M., Lippi, M., Maggini, M., Melacci, S.: Semantic video labeling by developmental visual agents. Comput. Vis. Image Underst. **146**, 9–26 (2016)

43. Gori, M., Lippi, M., Maggini, M., Melacci, S., Pelillo, M.: En plein air visual agents. In: Murino , V., Puppo, E., (eds.) Proceedings of Image Analysis and Processing—ICIAP 2015—18th International Conference, Genoa, Italy, September 7—11, 2015, , Part II, volume 9280 of Lecture Notes in Computer Science, pp. 697–709. Springer, Berlin (2015)

44. Hadizadeh, H., Bajić, I.V.: Saliency-aware video compression. IEEE Trans. Image Process. **23**(1), 19–33 (2013)

45. Hassanin, M., Khan, S., Tahtali, M.: Visual affordance and function understanding: a survey. CoRR (2018). abs/1807.06775

46. Hinton, G.E.: Learning distributed representations of concepts. In: Proceedings of the Eighth Annual Conference of the Cognitive Science Society, Amherst 1986, pp. 1–12. Lawrence Erlbaum, Hillsdale (1986)

47. Hinton, G.E., Sejnowski, T.J.: Learning and relearning in Boltzmann machines. In: Rumelhart, D.E., McClelland, J.L. (eds.) Parallel Distributed Processing, vol. 1, Chap. 7, pp. 282–317. MIT Press, Cambridge (1986)

48. Horn, B.K.P., Schunck, B.G.: Determining optical flow. Artif. Intell. **17**(1–3), 185–203 (1981)

49. Hubel, D.H., Wiesel, T.N.: Receptive fields, binocular interaction, and functional architecture in the cat's visual cortex. J. Physiol. (Lond.) **160**, 106–154 (1962)

50. Itti, L.: Automatic foveation for video compression using a neurobiological model of visual attention. IEEE Trans. Image Process. **13**(10), 1304–1318 (2004)

51. Itti, L., Koch, C., Niebur, E.: A model of saliency-based visual attention for rapid scene analysis. IEEE Trans. Pattern Anal. Mach. Intell. **11**, 1254–1259 (1998)

52. John David, J.: Classical electrodynamics. Wiley (2007)

53. Jiang, M., Boix, X., Roig, G., Juan, X., Van Gool, L., Zhao, Qi.: Learning to predict sequences of human visual fixations. IEEE Trans. Neural Netw. Learn. Syst. **27**(6), 1241–1252 (2016)

54. Khosla, D., Moore, C.K., Huber, D., Chelian, S.: Bio-inspired visual attention and object recognition. In: Intelligent Computing: Theory and Applications V, vol. 6560, pp. 656003. International Society for Optics and Photonics (2007)

55. Kim, B., Reif, E., Wattenberg, M., Bengio, S.: Do neural networks show gestalt phenomena? An exploration of the law of closure. CoRR (2019). abs/1903.01069

56. Kingstone, A., Daniel, S., Eastwood John, D.: Cognitive ethology: a new approach for studying human cognition. Br. J. Psychol. **99**, 317–340 (2010)

57. Koch. C., Ullman, S.: Shifts in selective visual attention: towards the underlying neural circuitry. In: Matters of Intelligence, pages 115–141. Springer, Berlin (1987)

58. Koch, K., McLean, J., Segev, R., Freed, M.A., Berry II, M.J., Balasubramanian, V., Sterling, P.: How much the eye tells the brain. Curr. Biol. **16**(14), 1428–1434 (2006)

59. Krizhevsky, A., Sutskever, I., Hinton, G.E.: Imagenet classification with deep convolutional neural networks. In: Pereira, F., Burges, C.J.C., Bottou, L., Weinberger, K.Q. (eds.) Advances in neural information processing systems, vol. 25, pp. 1097–1105. Curran Associates, Inc. (2012)

60. Kuang, X., Gibson, M., Shi, B.E., Rucci, M.: Active vision during coordinated head/eye movements in a humanoid robot. IEEE Trans. Robot. **28**(6) (2012)

61. Le Meur, O., Liu, Z.: Saccadic model of eye movements for free-viewing condition. Vis. Res. **116**, 152–164 (2015)

62. LeCun, Y., Bengio, Y., Hinton, G.: Deep learning. Nature **521**(7553), 436–444 (2015)

63. Lee, H., Grosse, R., Ranganath, R., Ng, A.Y.: Convolutional deep belief networks for scalable unsupervised learning of hierarchical representations. In: Proceedings of the 26th Annual International Conference on Machine Learning, ICML '09, pp. 609–616, New York. ACM (2009)

64. Lettvin, J.Y., Maturana, H.R., McCulloch, W.S., Pitts, W.H.: What the frog's eye tells the frog's brain. Proc. IRE **47**(11), 1940–1951 (1959)

65. Liero, M., Stefanelli, U.: A new minimum principle for Lagrangian mechanics. J. Nonlinear Sci. **23**(2), 179–204 (2013)

66. Liu, C., Mao, J., Sha, F., Yuille, A.: Attention correctness in neural image captioning. In: Thirty-First AAAI Conference on Artificial Intelligence (2017)

67. Lowe, D.G.: Distinctive image features from scale-invariant keypoints. Int. J. Comput. Vis. **60**(2), 91–110 (2004)
68. Bruce D Lucas, Takeo Kanade, et al. An iterative image registration technique with an application to stereo vision. Vancouver, British Columbia, 1981
69. Lupyan, G.: Linguistically modulated perception and cognition: the label-feedback hypothesis. Front. Psychol. **3** (2012)
70. Marinai, S., Gori, M., Soda, G.: Artificial neural networks for document analysis and recognition. IEEE Trans. Pattern Anal. Mach. Intell. **27**(1), 23–35 (2005)
71. Marr, D.: Vision: A Computational Investigation into the Human Representation and Processing of Visual Information. MIT press (2010)
72. Marr, D., Poggio, T.: From understanding computation to understanding neural circuitry (1976)
73. McAlonan, K., Cavanaugh, J., Wurtz, R.H.: Guarding the gateway to cortex with attention in visual thalamus. Nature **456**(7220), 391–394 (2008)
74. Meloni, E., Pasqualini, L., Tiezzi, M., Gori, M., Melacci, S.: Learning in virtual visual environments made simple. CoRR (2020). abs/2007.08224
75. Ott, J., Linstead, E., LaHaye, N., Baldi, P.: Learning in the machine: to share or not to share? Neural Netw. **126**, 235–249 (2020)
76. Poggio, T.A., Anselmi, F.: Visual Cortex and Deep Networks: Learning Invariant Representations, , 1st edn. The MIT Press (2016)
77. Rajesh, R.P.N., Ballard, D.H.: Predictive coding in the visual cortex: a functional interpretation of some extra-classical receptive-field effects. Nat. Neurosci. **2**, 79–87 (1999)
78. Ross, C.F.: The Tarsier Fovea: Functionless Vestige or Nocturnal Adaptation? pp. 477–537. Springer US, Boston (2004)
79. Schlingensiepen, K.-H., Campbell, F.W., Legge, G.E., Walker, T.D.: The importance of eye movements in the analysis of simple patterns. Vis. Res. **26**(7), 1111–1117 (1986)
80. Simonyan, K., Zisserman, A.: Very deep convolutional networks for large-scale image recognition. CoRR (2014). abs/1409.1556
81. Sohn, E.: The eyes of mammals reveal a dark past. Nature (2019)
82. Treisman, A.M.: Strategies and models of selective attention. Psychol. Rev. **76**(3), 282 (1969)
83. Treisman, A.M., Gelade, G.: A feature-integration theory of attention. Cogn. Psychol. **12**(1), 97–136 (1980)
84. Ullman, S.: The Interpretation of Visual Motion. The MIT Press Series in Artificial Intelligence, The MIT Press (1979)
85. Kastner, S., Ungerleider, L.G.: Mechanisms of visual attention in the human cortex. Ann. Rev. Neurosci. **23**(1), 315–341 (2000)
86. Walls, G.L.: The vertebrate eye and its adaptive radiation (1942)
87. Watanabe, S.: Pattern Recognition: Human and Mechanical. Wiley, USA (1985)
88. Wood, J.N.: A smoothness constraint on the development of object recognition. Cognition **153**, 140–145 (2016)
89. Xia, X., Kulis, B.: W-net: a deep model for fully unsupervised image segmentation (2017). arXiv preprint arXiv:1711.08506
90. Dario, Z., Gori, M.: Variational laws of visual attention for dynamic scenes. In: Advances in Neural Information Processing Systems, pp. 3823–3832 (2017)
91. Zanca, D., Gori, M., Melacci, S., Rufa, A.: Gravitational models explain shifts on human visual attention. Sci. Rep. **10**(1), 1–9 (2020)
92. Zanca, D., Melacci, S., Gori, M.: Gravitational laws of focus of attention. EEE Trans. Pattern Anal. Mach. Intell. **42**(12), 2983–2995 (2019)
93. Zhang, H.-B., Zhang, Y.-X., Zhong, B., Lei, Q., Yang, L., Du, J.-X., Chen, D.-S.: A comprehensive survey of vision-based human action recognition methods. Sensors **19**(5) (2019)
94. Zhou, Y., Dong, H., El Saddik, A.: Deep learning in next-frame prediction: a benchmark review. In: IEEE Access (2020)
95. Lawrence Zitnick, C., Dollár, P.: Edge boxes: locating object proposals from edges. In: Proceedings of Computer Vision—ECCV 2014—13th European Conference, Zurich, Switzerland, September 6–12, 2014, Part V, pp. 391–405 (2014)

Index

Printed in the United States
by Baker & Taylor Publisher Services